T0281352

SpringerBriefs in Physics

Series editors

Egor Babaev, University of Massachusetts, Massachusetts, USA
Malcolm Bremer, University of Bristol, Bristol, UK
Xavier Calmet, University of Sussex, Brighton, UK
Francesca Di Lodovico, Queen Mary University of London, London, UK
Pablo D. Esquinazi, University of Leipzig, Leipzig, Germany
Maarten Hoogerland, Universiy of Auckland, Auckland, New Zealand
Eric Le Ru, Victoria University of Wellington, Kelburn, New Zealand
Hans-Joachim Lewerenz, California Institute of Technology, Pasadena, USA
James Overduin, Towson University, Towson, USA
Vesselin Petkov, Concordia University, Montreal, Canada
Charles H.-T. Wang, University of Aberdeen, Aberdeen, UK
Andrew Whitaker, Queen's University Belfast, Belfast, UK
Stefan Theisen, Max-Planck-Institut für Gravitationsphysik, Golm, Germany

More information about this series at http://www.springer.com/series/8902

Vicente Cortés · Alexander S. Haupt

Mathematical Methods
of Classical Physics

 Springer

Vicente Cortés
Department of Mathematics and Center
 for Mathematical Physics
University of Hamburg
Hamburg
Germany

Alexander S. Haupt
Department of Mathematics and Center
 for Mathematical Physics
University of Hamburg
Hamburg
Germany

ISSN 2191-5423   ISSN 2191-5431   (electronic)
SpringerBriefs in Physics
ISBN 978-3-319-56462-3   ISBN 978-3-319-56463-0   (eBook)
DOI 10.1007/978-3-319-56463-0

Library of Congress Control Number: 2017935828

Printed on acid-free paper

This Springer imprint is published by Springer Nature
The registered company is Springer International Publishing AG
The registered company address is: Gewerbestrasse 11, 6330 Cham, Switzerland

Acknowledgements

We are very grateful to David Lindemann for careful proofreading of earlier versions of the manuscript of this book and for numerous constructive comments which helped improve the presentation of this book. We thank Thomas Leistner for drawing our attention to the Ref. [6] and Thomas Mohaupt for useful remarks concerning Chap. 1.

Contents

Symbols

$\mathscr{L}_2$	Linearized Lagrangian, 31
$\mathbf{L}$	Angular momentum vector, 14
M	Smooth manifold (configuration space), 5
m	Mass of a point particle, 6
n	Dimension of M, 8
$\mathrm{pr}^{(k)}Z$	k-th prolongation of Z, 54
$\mathbf{p}$	Momentum vector, 13
Q	(Noether) charge, 60
$(q^1, \ldots, q^n, \hat{q}^1, \ldots, \hat{q}^n)$	Induced/canonical local coordinates on TM associated with local coordinates $(x^1, \ldots, x^n)$ on M, 7
$(q^1, \ldots, q^n, p_1, \ldots, p_n)$	Darboux coordinates, 20
Ric	Ricci curvature tensor, 72
S	Action, 6
$scal$	Scalar curvature of (M, g), 72
$S[f]$	Action functional of a classical field theory, 47
$\mathfrak{S}$	Source manifold (possibly with boundary), 45
t	Time, 5
$\mathscr{T}$	Target manifold, 47
TM	Tangent bundle of M, 5
T^μ_ν	Component of the energy–momentum tensor, 76
V	Potential, 5
V_{eff}	Effective potential, 15
$(x^1, \ldots, x^n)$	Local coordinates on an open subset $U \subset M$, 7
X_f	Hamiltonian vector field associated with a smooth function f, 20
$\mathfrak{X}(M)$	Set of all smooth vector fields on M, 11
X^{ver}	Vertical lift of X, 11
α	Euler–Lagrange one-form, 50
α_i	Component of the Euler–Lagrange one-form in some local coordinate system, 8
δ_{ij}	Kronecker delta. Its value is defined to be 1 if the indices are equal, and 0 otherwise, 33
γ	Smooth curve in M, 6
κ	Gravitational coupling constant, 72
λ	Liouville form, 20
Λ	Cosmological constant, 75
Ω	Hessian matrix of the potential V, 32
ω	Canonical symplectic form, 20
ω_0	Frequency of a small oscillation, 33
π	Canonical projection from TM to M, 7
$\tau(f)$	Tension of f, 62
τ_0	Period of a small oscillation, 33
ξ	Eigenvector of W, 35
$\langle \cdot, \cdot \rangle$	Euclidean scalar product on $\mathbb{R}^n$, 5

$*$	Hodge star operator, 65
Δ	Laplace operator, 63
$(\dot{\cdot})$	Time derivate, $\dot{f}(t) = f'(t), \ddot{f}(t) = f''(t)$, 2
∇	Covariant derivative or connection, 9
Σ	We use an adapted version of *Einstein's summation convention* throughout this book: Upper and lower indices appearing with the same symbol within a term are to be summed over. We write the symbol Σ to indicate whenever this convention is employed. Owing to the aforementioned convention, the summation indices can be (and are usually) omitted below the symbol Σ, 7
$(\cdot)^{\mathsf{T}}$	Matrix transposition, 34

Chapter 1
Introduction

Abstract We define the framework of classical physics as considered in the present book. We briefly point out its place in the history of physics and its relation to modern physics. As the prime example of a theory of classical physics we introduce Newtonian mechanics and discuss its limitations. This leads to and motivates the study of different formulations of classical mechanics, such as Lagrangian and Hamiltonian mechanics, which are the subjects of later chapters. Finally, we explain why in this book, we take a mathematical perspective on central topics of classical physics.

Classical physics refers to the collection of physical theories that do not use quantum theory and often predate modern quantum physics. They can be traced back to Newton (17th century) and in some sense even further all the way to Aristotle, Archimedes, and other Greek philosophers of antiquity (starting in the 4th century BC). However, this does by far not mean that theories of classical physics are exclusively a subject of the past. They continue to play important roles in modern physics, for example in the study of macroscopic systems, such as fluids and planetary motions, where the effects of the quantum behavior of the microscopic constituents are irrelevant.

Newtonian mechanics is arguably the first mathematically rigorous and self-contained theory of classical physics. In its traditional formulation, Newton's theory comprises three physical laws known as Newton's laws of motion, describing the relationship between a body (usually assumed to be a point particle of constant mass $m > 0$) and forces acting upon it. They also quantify the resulting motion of the body in response to those forces and can be summarised, in an inertial reference frame, as follows.

First law: A body at rest will stay at rest and a body in uniform motion will stay in motion at constant velocity, unless acted upon by a net force.

Second law: The force F acting on a body is equal to the mass of the body times the acceleration a of the body:

$$F = m\,a \ .$$

$$(1.1)$$

Third law: For every action force there is a corresponding reaction force which is equal in magnitude and opposite in direction. This is often abbreviated as *actio = reactio*.

© The Author(s) 2017
V. Cortés and A.S. Haupt, *Mathematical Methods of Classical Physics*,
SpringerBriefs in Physics, DOI 10.1007/978-3-319-56463-0_1

Considering a point particle moving in $\mathbb{R}^3$, its acceleration at time t is given by the second time derivate of its position vector, denoted by $a(t) = \ddot{x}$. Throughout the entire book, we work in the C^∞-category, unless stated otherwise, that is, manifolds and maps between them are usually assumed to be smooth. The force F is generally allowed to depend on position x, velocity $\dot{x}$, and time t, that is $F = F(x, \dot{x}, t)$. Equation (1.1) then turns into

$$\ddot{x} = \frac{1}{m} F(x, \dot{x}, t) \,. \tag{1.2}$$

For m and $F(x, \dot{x}, t)$ given, this is a set of second-order ordinary differential equations known as *Newton's equations of motion*. Note that $F = 0$ if and only if the motion $t \to x(t)$ is linear and therefore Newton's first law is a special case of the second law. Also, the third law is a consequence of the second law in combination with conservation of momentum, which ultimately follows from translational invariance (see Corollary 2.15). Despite this redundancy in Newton's laws, all three highlight different aspects of important concepts of modern physics, namely the notion of inertial frame in Einstein's theory of relativity and the relation between different concepts of mass (inertial vs. gravitational) in gravitational theories (see, for example, [5]).

We remark that Eq. (1.1) is also sometimes written as $F = \dot{p}$, where $p = m\dot{x}$ is the *momentum* of the particle. This allows for the consideration of bodies with non-constant mass, such as rockets consuming their fuel, where however the second law needs to be applied to the total system including the lost (or gained) mass.

In order to solve Eq. (1.2) for a given mechanical system, one first needs to determine an actual expression for the net force $F(x, \dot{x}, t)$. However, this can be intricate in practice, especially when constraint forces are present. Typically for constraint forces it is easy to describe the constraints geometrically, while it is difficult to explicitly obtain the corresponding function $F(x, \dot{x}, t)$. For example, consider a ball moving on the surface of a flat table. Geometrically, the constraint imposed by the presence of the table implies that the vertical component of x (usually denoted by x^3) is held fixed. On the other hand, incorporating this constraint into Eq. (1.2) amounts to specifying an actual functional expression for the force exerted by the table on the ball. Another problem with Newton's formulation, partly related to the previous issue, is the intrinsic use of Cartesian coordinates. Indeed, changing to a different coordinate system is generally cumbersome.

Given these limitations, one is led to consider more geometric formulations, such as Lagrangian and Hamiltonian mechanics, in which generalized coordinates are introduced. These two formulations were introduced by Lagrange in 1788 and by Hamilton in 1833, respectively. They can be adapted to the mechanical system at hand, for example in order to incorporate geometrically the presence of constraint forces. In addition, symmetries can be identified more straightforwardly and exploited more efficiently in these alternative formulations of classical mechanics. Historically, in particular Hamilton–Jacobi theory – yet another formulation – has also played an important role in the development of quantum mechanics [3, 7, 18]. It has also paved the way for the development of classical field theory, which incorporates Einstein's theory of relativity and nowadays underlies many areas of modern physics.

For these reasons, a profound understanding of these central areas of classical physics is very important and hence, many textbooks on these subjects exist, such as for example Refs. [1, 2, 7, 8]. However, many of these textbooks require a strong background in physics and do not put strong emphasis on mathematical rigor. This presents difficulties for the more mathematically inclined reader.

As a complementary contribution to the existing literature, the present book takes a different, namely more mathematical, perspective at these central topics of classical physics. It puts emphasis on a mathematically precise formulation of the topics while conveying the underlying geometrical ideas. For this purpose the mathematical presentation style (definition, theorem, proof) is used and all theorems are proved. In addition, the field theory part is formulated in terms of the theory of jet bundles, highlighting the relativistic covariance. Jet bundles do not receive widespread attention in the physics literature up to now. Each chapter of the book is accompanied by a number of exercises, which can be found collectively in Appendix A. The interested reader is highly encouraged to try the exercises, as this will greatly help in gaining a deeper understanding of the subject.

This book grew out of a lecture course on "mathematical methods of classical physics" held in the winter semester of 2015/2016 as part of the master's program in mathematics and mathematical physics at the Department of Mathematics at the University of Hamburg. It is therefore primarily directed at readers with a background in mathematical physics and mathematics. Also, physicists with a strong interest in mathematics may find this text useful as a complementary resource.

Chapter 2
Lagrangian Mechanics

Abstract In this chapter, we lay out the foundations of Lagrangian Mechanics. We introduce the basic concepts of Lagrangian mechanical systems, namely the Lagrangian, the action, and the equations of motion, also known as the Euler–Lagrange equations. We also discuss important examples, such as the free particle, the harmonic oscillator, as well as motions in central force potentials, such as Newton's theory of gravity and Coulomb's electrostatic theory. Highlighting the importance of symmetries, we study integrals of motion and Noether's theorem. As an application, we consider motions in radial potentials and, further specializing to motions in Newton's gravitational potential, we conclude this section with a derivation of Kepler's laws of planetary motion.

2.1 Lagrangian Mechanical Systems and their Equations of Motion

Definition 2.1 A *Lagrangian mechanical system* is a pair $(M, \mathscr{L})$ consisting of a smooth manifold M and a smooth function $\mathscr{L}$ on TM. The manifold M is called the *configuration space* and the function $\mathscr{L}$ is called the *Lagrangian function* (or simply the *Lagrangian*) of the system.

More generally, one can allow for the Lagrangian to depend explicitly on an extra variable interpreted as time.

Example 2.2 The Lagrangian of a point particle of *mass* $m > 0$ moving in Euclidean space $\mathbb{R}^n$ under the influence of a *potential* $V \in C^\infty(\mathbb{R}^n)$ is

$$\mathscr{L}(v) = \frac{1}{2}m\langle v, v\rangle - V(x), \quad v \in T_x\mathbb{R}^n, \quad x \in \mathbb{R}^n,$$

where $\langle \cdot, \cdot \rangle$ denotes the Euclidean scalar product. Given a smooth curve $t \mapsto \gamma(t)$ in $\mathbb{R}^n$, the quantities $E_{\text{kin}}(t) = \frac{1}{2}m\langle \gamma'(t), \gamma'(t)\rangle$ and $E_{\text{pot}}(t) = V(\gamma(t))$ are called the *kinetic energy* and the *potential energy* at time t, respectively. Their sum

© The Author(s) 2017
V. Cortés and A.S. Haupt, *Mathematical Methods of Classical Physics*,
SpringerBriefs in Physics, DOI 10.1007/978-3-319-56463-0_2

$$E(t) = E_{\text{kin}}(t) + E_{\text{pot}}(t)$$

is the *total energy* at time t.

By specializing V one obtains many important mechanical systems:

1. The *free particle*: $V = 0$.
2. The *harmonic oscillator*: $n = 1$ and $V = \frac{1}{2}kx^2$, where k is a positive constant known as *Hooke's constant*.
3. *Newton's theory of gravity*: $n = 3$ and

$$V = -\frac{\kappa m M}{r},$$

where $r = |x|$ denotes the Euclidean norm of the vector x, $M > 0$ is the mass of the particle (placed at the origin $0 \in \mathbb{R}^3$) generating the gravitational potential, and κ is a positive constant known as *Newton's constant*. (Notice that in this case V is not defined at the origin, so the configuration space is $\mathbb{R}^3 \setminus \{0\}$ rather than $\mathbb{R}^3$.)
4. *Coulomb's electrostatic theory*: $n = 3$ and

$$V = \frac{k_e q_1 q_2}{r},$$

where q_1 is the charge of the particle of mass m moving under the influence of the electric potential generated by a particle of charge q_2 and k_e is a positive constant known as *Coulomb's constant*. Notice that up to a constant factor Coulomb's potential is of the same type as Newton's potential. However, contrary to Newton's potential, the Coulomb potential can have either sign. As we shall see, this corresponds to the fact that electric forces can be attractive or repulsive, depending on the sign of the product $q_1 q_2$, whereas gravitational forces are always attractive. Another important difference is that the gravitational potential contains the mass m as a factor whereas the electric potential does not. As we shall see, the former property implies that the acceleration γ'' of the particle in Newton's theory of gravity is independent of its mass.

Example 2.3 More generally, given a smooth function V on a pseudo-Riemannian manifold (M, g), one can consider the Lagrangian

$$\mathscr{L}(v) = \frac{1}{2} g(v, v) - V(x), \quad v \in T_x M, \quad x \in M.$$

Definition 2.4 Let $(M, \mathscr{L})$ be a Lagrangian mechanical system. The *action* of a smooth curve $\gamma : [a, b] \to M$ is defined as

$$S(\gamma) := \int_a^b \mathscr{L}(\gamma'(t)) dt.$$

A *motion* of the system is a critical point of S under smooth variations with fixed endpoints. (This statement is a mathematical formulation of *Hamilton's principle of least action*, which should better be called principle of stationary action.)

Next we will derive the *equations of motion*, which are the differential equations describing critical points of the action functional. These are known as the *Euler–Lagrange equations* and are given in (2.3). For this we introduce coordinates $(x^1, \ldots, x^n)$ on an open subset $U \subset M$. They induce a system of coordinates $(q, \hat{q}) = (q^1, \ldots, q^n, \hat{q}^1, \ldots, \hat{q}^n)$ on the open subset $TU = \pi^{-1}(U) \subset TM$, where $\pi : TM \to M$ denotes the canonical projection. The *induced coordinates* are defined by

$$q^i := x^i \circ \pi, \quad \hat{q}^i(v) := dx^i(v), \quad v \in TU.$$

Let $\gamma_s : [a, b] \to M$, $-\varepsilon < s < \varepsilon$, $\varepsilon > 0$, be a smooth variation of $\gamma : [a, b] \to M$ with fixed endpoints. Then in local coordinates as above we compute

$$\frac{d}{ds}\bigg|_{s=0} \mathcal{L}(\gamma_s') = \sum \left(\frac{\partial \mathcal{L}}{\partial q^i}(\gamma') \frac{d}{ds}\bigg|_{s=0} q^i(\gamma_s') + \frac{\partial \mathcal{L}}{\partial \hat{q}^i}(\gamma') \frac{d}{ds}\bigg|_{s=0} \hat{q}^i(\gamma_s') \right)$$
$$= \sum \left(\frac{\partial \mathcal{L}}{\partial q^i}(\gamma') \frac{d}{ds}\bigg|_{s=0} \gamma_s^i + \frac{\partial \mathcal{L}}{\partial \hat{q}^i}(\gamma') \frac{d}{ds}\bigg|_{s=0} \dot{\gamma}_s^i \right),$$

where $\gamma^i := x^i \circ \gamma$ and $\dot{\gamma}^i := (\gamma^i)'$. Note that here and throughout the rest of the book, we use an adapted version of *Einstein's summation convention*, where repeated upper and lower indices are to be summed over. This is indicated by a plain sum symbol (see also p. xi). If we denote by $\mathcal{V} := \frac{d}{ds}\big|_{s=0} \gamma_s$ the variation vector field along γ, then we can rewrite this as

$$\frac{d}{ds}\bigg|_{s=0} \mathcal{L}(\gamma_s') = \sum \left(\frac{\partial \mathcal{L}}{\partial q^i}(\gamma')\mathcal{V}^i + \frac{\partial \mathcal{L}}{\partial \hat{q}^i}(\gamma')\dot{\mathcal{V}}^i \right)$$
$$= \sum \left(\frac{\partial \mathcal{L}}{\partial q^i}(\gamma') - \frac{d}{dt}\frac{\partial \mathcal{L}}{\partial \hat{q}^i}(\gamma') \right) \mathcal{V}^i + f',$$

where $\mathcal{V}^i := dx^i(\mathcal{V}) = \hat{q}^i(\mathcal{V})$ are the components of $\mathcal{V}$ and

$$f := \sum \frac{\partial \mathcal{L}}{\partial \hat{q}^i}(\gamma')\mathcal{V}^i.$$

Notice that the local vector field $\mathcal{V}^{\mathrm{ver}} = \sum \mathcal{V}^i \frac{\partial}{\partial \hat{q}^i}$ along γ' is independent of the choice of local coordinates and, hence, defines a global vector field along γ'. This follows from the fact that it corresponds to $\mathcal{V}$ under the canonical identification of $T_{\gamma(t)}M$ with the vertical tangent space

$$T_{\gamma'(t)}^{\mathrm{ver}}(TM) := \ker d\pi|_{\gamma'(t)} \subset T_{\gamma'(t)}(TM).$$

Using this vector field we can rewrite the function f as

$$f = \mathscr{V}^{\text{ver}}(\mathscr{L})|_{\gamma'}.$$

This shows that also f is globally defined on the interval $[a, b]$ and vanishes at the endpoints. Therefore we have

$$\left.\frac{d}{ds}\right|_{s=0} S(\gamma_s) = \left.\frac{d}{ds}\right|_{s=0} \int_a^b \mathscr{L}(\gamma_s'(t))dt = \int_a^b \left.\frac{d}{ds}\right|_{s=0} \mathscr{L}(\gamma_s'(t))dt \quad (2.1)$$

$$= \int_a^b \left(\left.\frac{d}{ds}\right|_{s=0} \mathscr{L}(\gamma_s'(t)) - f'(t)\right) dt, \quad (2.2)$$

where the integrand is a globally defined function on the interval $[a, b]$ with the following local expression:

$$\left.\frac{d}{ds}\right|_{s=0} \mathscr{L}(\gamma_s'(t)) - f'(t) = \sum \left(\frac{\partial \mathscr{L}}{\partial q^i}(\gamma'(t)) - \frac{d}{dt}\frac{\partial \mathscr{L}}{\partial \hat{q}^i}(\gamma'(t))\right) \mathscr{V}^i(t).$$

Theorem 2.5 *Let $(M, \mathscr{L})$ be a Lagrangian mechanical system, $n = \dim M$. The motions of the system are the solutions $\gamma : [a, b] \to M$ of the following system of ordinary differential equations:*

$$\alpha_i := \frac{\partial \mathscr{L}}{\partial q^i}(\gamma') - \frac{d}{dt}\frac{\partial \mathscr{L}}{\partial \hat{q}^i}(\gamma') = 0, \quad i = 1, \ldots n. \quad (2.3)$$

Proof Let us first remark that for every vector field $\mathscr{V}$ along γ the function $\sum \alpha_i \mathscr{V}^i = \left.\frac{d}{ds}\right|_{s=0} \mathscr{L}(\gamma_s') - f'$ is coordinate independent and thus globally well-defined. From this we can deduce that $\alpha = \sum \alpha_i dx^i|_\gamma$ is a well-defined one-form along γ (cf. Appendix A, Exercise 1). In virtue of (2.2), we see that γ is a motion if and only if

$$\int_a^b \alpha(\mathscr{V})|_t dt = 0.$$

for all vector fields $\mathscr{V}$ along γ vanishing at the endpoints. To see that this implies that $\alpha(t_0) = 0$ for all $t_0 \in (a, b)$ we take $\delta > 0$ such that $\gamma([t_0 - \delta, t_0 + \delta])$ is contained in a coordinate domain and consider $\mathscr{V}^i = h\alpha_i$, where $h \geq 0$ is a smooth function on $[a, b]$ with support contained in $(t_0 - \delta, t_0 + \delta)$ such that $h(t_0) > 0$. This defines a smooth vector field $\mathscr{V}$ along γ vanishing at the endpoints and $0 = \int_a^b \alpha(\mathscr{V})|_t dt = \int_{t_0-\delta}^{t_0+\delta} h \sum \alpha_i^2 dt$ now implies that $\alpha(t_0) = 0$. This proves that $\alpha|_{(a,b)} = 0$ and, by continuity, $\alpha = 0$. $\qquad \square$

Definition 2.6 The one-form α along γ will be called the *Euler–Lagrange one-form*.

Example 2.7 Let (M, g) be a pseudo-Riemannian metric and consider the Lagrangian $\mathscr{L}(v) = \frac{1}{2}g(v, v)$, $v \in TM$. In canonical local coordinates $(q^1, \ldots, q^n, \hat{q}^1, \ldots, \hat{q}^n)$ on TM associated with local coordinates $(x^1, \ldots, x^n)$ on M it is given by

$$\mathscr{L} = \frac{1}{2}\sum \tilde{g}_{ij}\hat{q}^i\hat{q}^j,$$

where $g = \sum g_{ij}dx^idx^j$ and $\tilde{g}_{ij} = g_{ij} \circ \pi$. We compute

$$\frac{\partial \mathscr{L}}{\partial q^i} = \frac{1}{2}\sum \frac{\partial \tilde{g}_{kj}}{\partial q^i}\hat{q}^k\hat{q}^j = \frac{1}{2}\sum \left(\frac{\partial g_{kj}}{\partial x^i} \circ \pi\right)\hat{q}^k\hat{q}^j,$$

$$\frac{\partial \mathscr{L}}{\partial \hat{q}^i} = \sum \tilde{g}_{ij}\hat{q}^j,$$

$$\frac{\partial \mathscr{L}}{\partial q^i}(\gamma') = \frac{1}{2}\sum \frac{\partial g_{kj}}{\partial x^i}(\gamma)\dot{\gamma}^k\dot{\gamma}^j,$$

$$\frac{\partial \mathscr{L}}{\partial \hat{q}^i}(\gamma') = \sum g_{ij}(\gamma)\dot{\gamma}^j,$$

$$\frac{d}{dt}\frac{\partial \mathscr{L}}{\partial \hat{q}^i}(\gamma') = \sum \frac{\partial g_{ij}}{\partial x^k}(\gamma)\dot{\gamma}^k\dot{\gamma}^j + \sum g_{ij}(\gamma)\ddot{\gamma}^j,$$

$$-\alpha_i = \sum g_{ij}(\gamma)\ddot{\gamma}^j + \frac{1}{2}\sum \left(\frac{\partial g_{ij}}{\partial x^k}(\gamma) + \frac{\partial g_{ik}}{\partial x^j}(\gamma) - \frac{\partial g_{kj}}{\partial x^i}(\gamma)\right)\dot{\gamma}^k\dot{\gamma}^j$$

$$= \sum_\ell g_{i\ell}(\gamma)\left(\ddot{\gamma}^\ell + \sum_{j,k}\Gamma^\ell_{jk}(\gamma)\dot{\gamma}^j\dot{\gamma}^k\right).$$

This shows that the Euler–Lagrange equations are equivalent to the geodesic equations

$$\ddot{\gamma}^\ell + \sum \Gamma^\ell_{jk}(\gamma)\dot{\gamma}^j\dot{\gamma}^k = 0, \quad \ell = 1, \ldots, n.$$

Using the covariant derivative ∇ they can be written as $\nabla_{\gamma'}\gamma' = 0$ or $\frac{\nabla}{dt}\gamma' = 0$.

Proposition 2.8 *Let V be a smooth function on a pseudo-Riemannian manifold (M, g) and consider the Lagrangian*

$$\mathscr{L}(v) = \frac{1}{2}g(v, v) - V(\pi(v)), \quad v \in TM, \tag{2.4}$$

of Example 2.3. Then a curve $\gamma : I \to M$ is a motion if and only if it satisfies the following equation

$$\frac{\nabla}{dt}\gamma' + \text{grad } V|_\gamma = 0 \tag{2.5}$$

Proof We have already shown that the Euler–Lagrange one-form in the case $V = 0$ is given by $\alpha = -g\left(\frac{\nabla}{dt}\gamma', \cdot\right)$. It remains to compute the contribution of the potential V to the Euler–Lagrange one-form, which is

$$-\sum \frac{\partial(V \circ \pi)}{\partial q^i}(\gamma')dx^i = -\sum \frac{\partial V}{\partial x^i}(\gamma)dx^i = -dV_\gamma.$$

So in total we obtain

$$-\alpha = g\left(\frac{\nabla}{dt}\gamma', \cdot\right) + dV|_\gamma$$

and, hence,

$$-g^{-1}\alpha = \frac{\nabla}{dt}\gamma' + \operatorname{grad} V|_\gamma,$$

where $g^{-1} : T^*M \to TM$ denotes the inverse of the map

$$g : TM \to T^*M, \quad v \mapsto g(v, \cdot).$$

$\square$

2.2 Integrals of Motion

Definition 2.9 Let $(M, \mathscr{L})$ be a Lagrangian mechanical system. A smooth function f on TM is called an *integral of motion* if it is constant along every motion of the system, that is for every motion $\gamma : I \to M$ the function $f(\gamma')$ is constant.

Obviously, the integrals of motion are completely determined by the equations of motion and do not depend on the precise Lagrangian. Therefore, the notion of an integral of motion is meaningful if we are just given a system of second order differential equations[1] for a curve $\gamma : I \to M$ in a smooth manifold M.

Proposition 2.10 (Conservation of energy) *Let V be a smooth function on a pseudo-Riemannian manifold (M, g) and consider the Lagrangian*

$$\mathscr{L}(v) = \frac{1}{2}g(v, v) - V(\pi(v)), \quad v \in TM, \tag{2.6}$$

of Example 2.3. Then the total energy

$$v \mapsto E(v) := \frac{1}{2}g(v, v) + V(\pi(v))$$

---

[1] Such a system will be usually given by a consistent specification of a system of second order differential equations for the components of the curve in each local coordinate system. A typical example is (2.3).

is conserved, *that is it is an integral of motion. If V is constant then the motion is geodesic.*

Proof We compute the derivative of $t \mapsto E(\gamma'(t))$ along a motion γ:

$$\frac{d}{dt}\left(\frac{1}{2}g(\gamma', \gamma') + V(\gamma)\right) = g\left(\frac{\nabla}{dt}\gamma', \gamma'\right) + dV|_\gamma \gamma' \stackrel{(2.5)}{=} 0.$$

If V is constant then the equation of motion (2.5) reduces to the geodesic equation.
□

Remark The last statement of the proposition is related to conservation of momentum in Newtonian mechanics. Recall from Chap. 1 that the momentum of a particle $\gamma : I \to \mathbb{R}^n$ of mass m moving in Euclidean space is $\mathbf{p} = m\gamma'$. Newton's second law of mechanics has the form $\mathbf{p}' = F$, where F is the force acting on the particle, which may depend on the position γ and momentum $\mathbf{p}$ of the particle. Obviously, $F = 0$ implies that the momentum is constant along every motion. The conservation of energy holds in Newtonian mechanics if the force field F depends only on the position and is a gradient vector field. In fact, then we can write $F = -\text{grad } V$ for some function V on $\mathbb{R}^n$ and hence, Newton's equation is a special case of (2.5).

Definition 2.11 Let $(M, \mathscr{L})$ be a Lagrangian mechanical system. A diffeomorphism $\varphi : M \to M$ is called an *automorphism* of the system if

$$\mathscr{L}(d\varphi v) = \mathscr{L}(v),$$

for all $v \in TM$. A vector field $X \in \mathfrak{X}(M)$ is called an *infinitesimal automorphism* if its flow consists of local automorphisms of $(M, \mathscr{L})$.

Theorem 2.12 (Noether's theorem) *With every infinitesimal automorphism X of a Lagrangian mechanical system $(M, \mathscr{L})$ we can associate an integral of motion $f = d\mathscr{L}X^{ver}$, where $X^{ver} \in \mathfrak{X}(TM)$ denotes the vertical lift of X. (In local coordinates we have $X^{ver} = \sum(X^i \circ \pi)\frac{\partial}{\partial \hat{q}^i}$ if $X = \sum X^i \frac{\partial}{\partial x^i}$.)*

Proof Let us denote by φ_s the flow of X. For every motion $\gamma : I \to M$ of the system with values in the domain of definition of φ_s the curve $\gamma_s = \varphi_s \circ \gamma$ is again a motion and

$$\gamma'_s = d\varphi_s \gamma', \quad \frac{\partial}{\partial s}\bigg|_{s=0} \gamma'_s = \tilde{X}|_{\gamma'},$$

where

$$TM \ni v \mapsto \tilde{X}(v) = \frac{\partial}{\partial s}\bigg|_{s=0} d\varphi_s(v)$$

denotes the vector field on TM the flow of which is $d\varphi_s$. Differentiating the equation $\pi \circ d\varphi_s = \varphi_s \circ \pi$ with respect to s, we obtain $d\pi \circ \tilde{X} = X \circ \pi$. In local coordinates (x^i) on M and corresponding local coordinates $(q^i, \hat{q}^i)$ on TM this means that

$$\tilde{X} = \sum \left((X^i \circ \pi)\frac{\partial}{\partial q^i} + Y^i \frac{\partial}{\partial \hat{q}^i} \right), \quad X = \sum X^i \frac{\partial}{\partial x^i}.$$

The locally defined functions Y^i on TM can be computed at $v \in TM$ as follows:

$$Y^i(v) = d\hat{q}^i \tilde{X}(v) = \frac{\partial}{\partial s}\bigg|_{s=0} \hat{q}^i(d\varphi_s v) = \frac{\partial}{\partial s}\bigg|_{s=0} \hat{q}^i \left(\sum_{j,k} v^j \frac{\partial \varphi_s^k}{\partial x^j} \frac{\partial}{\partial x^k} \right)$$

$$= \frac{\partial}{\partial s}\bigg|_{s=0} \sum_j v^j \frac{\partial \varphi_s^i}{\partial x^j} = dX^i(v),$$

where $\varphi_s^i := x^i \circ \varphi_s$ and we have used that $\frac{\partial}{\partial s}\big|_{s=0} \varphi_s^i = X^i$. The above implies that

$$0 = \frac{\partial}{\partial s}\bigg|_{s=0} \mathscr{L}(\gamma_s') = d\mathscr{L}\tilde{X}|_{\gamma'} = \sum \left(\frac{\partial \mathscr{L}}{\partial q^i}\bigg|_{\gamma'} X^i(\gamma) + \frac{\partial \mathscr{L}}{\partial \hat{q}^i}\bigg|_{\gamma'} dX^i(\gamma') \right).$$

$$(2.7)$$

Using the Euler–Lagrange equations we can now compute

$$\frac{d}{dt}f(\gamma') = \frac{d}{dt} \sum \frac{\partial \mathscr{L}}{\partial \hat{q}^i}\bigg|_{\gamma'} X^i(\gamma)$$

$$= \sum \underbrace{\left(\frac{d}{dt} \frac{\partial \mathscr{L}}{\partial \hat{q}^i}\bigg|_{\gamma'} \right)}_{= \frac{\partial \mathscr{L}}{\partial q^i}|_{\gamma'}} X^i(\gamma) + \sum \frac{\partial \mathscr{L}}{\partial \hat{q}^i}\bigg|_{\gamma'} dX^i(\gamma') \overset{(2.7)}{=} 0.$$

□

Proposition 2.13 *The group of automorphisms of the Lagrangian system of Example 2.3 is given by the Lie subgroup*

$$\mathrm{Aut}(M, \mathscr{L}) = \{\varphi \in \mathrm{Isom}(M, g) | V \circ \varphi = V\} \subset \mathrm{Isom}(M, g).$$

Its Lie algebra consists of all Killing vector fields X such that $X(V) = 0$.

Proof See Appendix A, Exercise 6.   □

Corollary 2.14 *Let V be a smooth function on a pseudo-Riemannian manifold (M, g) and consider the Lagrangian $\mathscr{L}(v) = \frac{1}{2}g(v, v) - V(\pi(v))$, $v \in TM$, of Example 2.3. Then every Killing vector field X such that $X(V) = 0$ gives rise to an integral of motion $f(v) = g(v, X(\pi(v)))$, $v \in TM$.*

Corollary 2.15 (Conservation of momentum) *Consider, as in the previous corollary, the Lagrangian of Example 2.3. Assume that with respect to some coordinate system* *on M the metric g and the potential V are both invariant under translations in one of the coordinates x^i. Then the function*

$$p_i := \sum \tilde{g}_{ij} \hat{q}^j$$

is an integral of motion on the coordinate domain.

Since the Euclidian metric is translational invariant we have the following special case of Corollary 2.15.

Corollary 2.16 *Let $V = V(x^2, \ldots, x^n)$ be a smooth function on Euclidean space $\mathbb{R}^n$ which does not depend on the first coordinate and consider the Lagrangian $\mathscr{L}(v) = \frac{1}{2}m\langle v, v\rangle - V(\pi(v))$, $v \in T\mathbb{R}^n$, of Example 2.2. Then the first component p_1 of the momentum vector $\mathbf{p} = \sum p_i e_i$ is an integral of motion, where $p_i(v) := m\langle v, e_i\rangle$, $v \in T\mathbb{R}^n$.*

Note that the result still holds if we replace the Euclidean scalar product by a pseudo-Euclidean scalar product, such as the Minkowski scalar product.

Corollary 2.17 (Conservation of angular momentum) *Let V be a smooth radial function on $\mathbb{R}^3 \setminus \{0\}$, that is V depends only on the radial coordinate r, and consider the Lagrangian $\mathscr{L}(v) = \frac{1}{2}m\langle v, v\rangle - V(\pi(v))$, $v \in T(\mathbb{R}^3 \setminus \{0\})$. Then the components of the angular momentum vector*

$$\mathbf{L}(v) = \pi(v) \times \mathbf{p}(v) = m \begin{pmatrix} x^1 \\ x^2 \\ x^3 \end{pmatrix} \times \begin{pmatrix} v^1 \\ v^2 \\ v^3 \end{pmatrix}$$

are integrals of motion. Here, x^1, x^2, x^3 denote the components of $x = \pi(v)$.

Proof This can be proven either directly from the equations of motion or by applying Noether's theorem (see Appendix A, Exercise 9). The first proof uses the fact that the moment of force $\gamma \times F(\gamma)$, $F = -\text{grad } V$, is zero for every curve $\gamma : I \to \mathbb{R}^3 \setminus \{0\}$ and yields also the following proposition. $\qquad\square$

Definition 2.18 A vector field F on $\mathbb{R}^3 \setminus \{0\}$ is called *radial* if there exists a radial function f on $\mathbb{R}^3 \setminus \{0\}$ such that $F(x) = f(x)\frac{x}{|x|}$ for all $x \in \mathbb{R}^3 \setminus \{0\}$.

Proposition 2.19 *Let F be a smooth radial vector field on $\mathbb{R}^3 \setminus \{0\}$. Then the angular momentum is constant for every solution of Newton's equation $\frac{d}{dt}\mathbf{p}(t) = F(t)$, where we are using the usual notation $\mathbf{p}(t) = \mathbf{p}(\gamma'(t))$ and $F(t) = F(\gamma(t))$.*

Remark A function or vector field on $\mathbb{R}^3 \setminus \{0\}$ is radial if and only if it is *spherically symmetric*, that is invariant under SO(3) (or, equivalently, O(3)), see Appendix A, Exercises 11 and 12. The name is due to the fact that SO(3) is the group of orientation preserving isometries of the sphere S^2 and $O(3) = \text{Isom}(S^2)$.

In the next section we will see how to use the conservation of energy and angular momentum to analyze the motion in a radial potential.

2.3   Motion in a Radial Potential

We consider the Lagrangian $\mathscr{L}(v) = \frac{1}{2}m\langle v, v\rangle - V(r)$ on $\mathbb{R}^3 \setminus \{0\}$, where V is a smooth function of the radial coordinate $r = |x|$ alone. Since the mass can be absorbed into the definitions of V and $\mathscr{L}$ (denoting $\mathscr{L}/m$ and V/m again by $\mathscr{L}$ and V) we may as well put $m = 1$. So from now on we consider a particle of unit mass.

We know by the results of the previous section that the energy $E(t) = \frac{1}{2}\langle \gamma'(t), \gamma'(t)\rangle + V(|\gamma(t)|)$ and angular momentum $\mathbf{L}(t) = \gamma(t) \times \gamma'(t)$ are constant for every motion $\gamma : I \to \mathbb{R}^3 \setminus \{0\}$ of the system.

The first observation is that the conservation of angular momentum implies that the motion is *planar*, that is contained in a plane.

Proposition 2.20 *If the vector* $\mathbf{L} = \gamma \times \gamma'$ *is constant along the curve* $\gamma : I \to \mathbb{R}^3 \setminus \{0\}$ *then* γ *is a planar curve.*

Proof If $\mathbf{L} = 0$ then γ is a radial curve (see Appendix A, Exercise 7) and thus planar. Therefore we can assume that $\mathbf{L}$ is a nonzero constant vector. Since the cross product of two vectors is always perpendicular to both of them, we know that $\gamma(t) \in \mathbf{L}^\perp$ for all t. So γ is contained in the plane $\mathbf{L}^\perp$. $\square$

By a change of coordinates we can assume that the motion is restricted to the plane $e_3^\perp$ and use polar coordinates (r, φ) in that plane rather than Cartesian coordinates (x^1, x^2) to describe the motion. In these coordinates, the Euclidean metric is $dr^2 + r^2 d\varphi^2$ and, hence,

$$E = \frac{1}{2}\left(\dot{r}^2 + r^2\dot{\varphi}^2\right) + V(r). \tag{2.8}$$

To compute the cross product $\gamma \times \gamma'$ in polar coordinates we recall that

$$\partial_r = \cos\varphi\,\partial_1 + \sin\varphi\,\partial_2, \quad \partial_\varphi = -r\sin\varphi\,\partial_1 + r\cos\varphi\,\partial_2,$$

where we have abbreviated $\partial_r = \partial/\partial r$, $\partial_\varphi = \partial/\partial\varphi$ and $\partial_i = \partial/\partial x^i$. From these relations we easily obtain

$$\partial_r \times \partial_\varphi = re_3$$

and hence

$$\mathbf{L} = \gamma \times \gamma' = r\partial_r \times \left(\dot{r}\partial_r + \dot{\varphi}\partial_\varphi\right) = r^2\dot{\varphi}e_3.$$

Changing the time parameter t to $-t$, if necessary, we can assume that the constant $r^2\dot{\varphi} \geq 0$ and, hence,

$$L := |\mathbf{L}| = r^2\dot{\varphi}. \tag{2.9}$$

Substituting $\dot{\varphi} = L/r^2$ into the energy equation (2.8) we arrive at

$$E = \frac{1}{2}\dot{r}^2 + V_{\text{eff}}(r), \quad V_{\text{eff}}(r) := V(r) + \frac{L^2}{2r^2}, \tag{2.10}$$

where V_{eff} is called the *effective potential*. As a consequence, we obtain the following result.

Theorem 2.21 *Consider a particle of unit mass moving in a radial potential $V(r)$ in $\mathbb{R}^3$. Then the radial coordinate obeys the equation of motion $\ddot{r} = -V'_{\text{eff}}(r)$ of a particle $t \mapsto r(t) > 0$ moving in the Euclidean line under the potential*

$$V_{\text{eff}}(r) := V(r) + \frac{L^2}{2r^2}, \tag{2.11}$$

which reduces to the first order equation

$$\dot{r} = \pm\sqrt{2\,(E - V_{\text{eff}}(r))} \tag{2.12}$$

solvable by separation of variables. The angular coordinate is then given by

$$\varphi(t) = L \int_0^t \frac{ds}{r(s)^2} + \varphi(0),$$

where the initial value $\varphi(0)$ is freely specifiable or, as a function of r, by

$$\varphi(r) = \pm L \int_{r_0}^r \frac{ds}{s^2\sqrt{2\,(E - V_{\text{eff}}(s))}} + \varphi(r_0).$$

Proof The energy equation (2.10) immediately implies (2.12), which can be solved by separation of variables:

$$\pm\frac{dr}{\sqrt{2\,(E - V_{\text{eff}}(r))}} = dt.$$

Integration yields t as a function of r. Then r as a function of t is the inverse of that function. Differentiating (2.12) gives Newton's equation $\ddot{r} = -V'_{\text{eff}}(r)$. Given r as a function of t, the angular coordinate is now determined from $\dot{\varphi} = L/r^2$ by integration as claimed. Finally, to express φ as a function of r one does not need to know the inverse of the function $r \mapsto t(r)$ but can proceed as follows:

$$\frac{d}{dr}\varphi = \dot{\varphi}\frac{dt}{dr} = \pm\frac{L}{r^2\sqrt{2\,(E - V_{\text{eff}}(r))}}.$$

$\square$

Next we specialize the discussion to Newton's potential $V(r) = -\frac{M}{r}$, from which we will derive, in particular, Kepler's laws of planetary motion.

2.3.1 Motion in Newton's Gravitational Potential

By (2.11) we know that the motion in Newton's gravitational potential is governed by the following effective potential

$$f(r) := V_{\text{eff}}(r) = -\frac{M}{r} + \frac{L^2}{2r^2}. \tag{2.13}$$

We will assume that $L > 0$; the case $L = 0$ is treated in Appendix A, Exercise 8. The qualitative behavior of the effective potential is essential for the qualitative analysis of the possible orbits. The proof of the next proposition is elementary.

Proposition 2.22 *The function $f : \mathbb{R}^{>0} \to \mathbb{R}$ defined in (2.13) has the following properties:*

1. *f has a unique zero, at $r_0 := \frac{L^2}{2M}$.*
2. *It is positive for $r < r_0$ and negative for $r > r_0$.*
3. *$\lim_{r \to 0} f(r) = \infty$, $\lim_{r \to \infty} f(r) = 0$.*
4. *f has a unique critical point, at $r_{\min} := \frac{L^2}{M} > r_0$, where f attains its global minimum $f(r_{\min}) = -\frac{M^2}{2L^2}$.*
5. *f is strictly decreasing for $r < r_{\min}$ and strictly increasing for $r > r_{\min}$.*

Proposition 2.23

(i) *The range of possible energies for motions with nonzero angular momentum in Newton's gravitational potential is given by the interval $\left[-\frac{M^2}{2L^2}, \infty\right)$. The motions are unbounded if $E \geq 0$ and bounded if $E < 0$. The orbit of minimal energy $E_{\min} = -\frac{M^2}{2L^2}$ is a circle of radius $\frac{L^2}{M}$.*

(ii) *The perihelion distance r_{per}, that is the minimal distance to the origin, for a given energy $E \geq 0$ (and $L > 0$) is given by the unique positive solution of the equation $r^2 + \frac{M}{E}r - \frac{L^2}{2E} = 0$, which is $r_{\text{per}} = \frac{M}{2E}\left(-1 + \sqrt{1 + 2\frac{L^2 E}{M^2}}\right)$. For $E_{\min} < E < 0$ (and $L > 0$) the distance to the origin varies between the two positive solutions of the equation $r^2 + \frac{M}{E}r - \frac{L^2}{2E} = 0$, which are $r_{\text{per}} = \frac{M}{2|E|}\left(1 - \sqrt{1 + 2\frac{L^2 E}{M^2}}\right)$ and $r_{\text{aph}} = \frac{M}{2|E|}\left(1 + \sqrt{1 + 2\frac{L^2 E}{M^2}}\right)$ (the aphelion distance, that is the maximal distance to the origin).*

Proof The first part follows immediately from Proposition 2.22. For the second part note that at the points of minimal or maximal distance to the origin we have $\dot{r} = 0$ and, hence, $E = V_{\text{eff}} = f$. The last equation is equivalent to the quadratic equation $r^2 + \frac{M}{E}r - \frac{L^2}{2E} = 0$, which has a unique solution for $E = E_{\min}$ (the radius of the circular orbit), two positive solutions for $E_{\min} < E < 0$, and only one positive solution for $E \geq 0$.   □

Next we will solve the equations of motion. The function $t \mapsto r(t)$ is determined by

$$\ddot{r} = -f'(r) = -\frac{M}{r^2} + \frac{L^2}{r^3}.$$

To solve this equation we make the substitution $u = 1/r$ and compute

$$\frac{du}{d\varphi} = \left.\frac{du}{dt}\right|_{t(\varphi)} \frac{dt}{d\varphi} = \left.-\frac{\dot{r}}{r^2 \dot{\varphi}}\right|_{t(\varphi)} = \left.-\frac{\dot{r}}{L}\right|_{t(\varphi)},$$

which implies

$$\frac{d^2 u}{d\varphi^2} = -\frac{\ddot{r}}{L}\frac{dt}{d\varphi} = -\frac{\ddot{r}}{L\dot{\varphi}} = -\frac{\ddot{r} r^2}{L^2} = \frac{f'(r) r^2}{L^2} = \frac{M}{L^2} - u.$$

So we obtain the equation

$$\frac{d^2 u}{d\varphi^2} + u = \frac{M}{L^2},$$

the general solution of which is

$$u = k\cos(\varphi - \varphi_0) + \frac{M}{L^2},$$

where $k \geq 0$ and φ_0 are constants. By choosing $\varphi_0 = 0$, this allows us to write

$$r = \frac{p}{1 + \varepsilon\cos\varphi}, \quad p := \frac{L^2}{M}, \quad \varepsilon := \frac{kL^2}{M}. \tag{2.14}$$

For $0 \leq \varepsilon < 1$ this is an ellipse of eccentricity ε.

For $\varepsilon = 1$ it is a parabola and for $\varepsilon > 1$ a component of a hyperbola. (Observe that for the parabola and hyperbola the angle φ is constrained by the condition that $\varepsilon\cos\varphi > -1$.) Recall that for an ellipse with major half-axis a and minor half-axis b the distance between the two focal points is $2c$, where $c^2 = a^2 - b^2$, and the eccentricity and parameter are given by

$$\varepsilon = \frac{c}{a}, \quad p = \frac{b^2}{a},$$

see e.g. [10, VIII.43] for a detailed discussion of conic sections.

Summarizing we obtain:

Theorem 2.24 *The motions of a particle (with nonzero angular momentum) in Newton's gravitational potential are along conic sections with a focus at the origin. The bounded motions are ellipses.*

The latter statement is the content of *Kepler's first law* (of planetary motion): The planets move along ellipses with the sun at a focal point. *Kepler's second law* asserts that the area swept out during a time interval $[t, t + s]$ by the line segment connecting the sun and the planet depends only on its duration s and not on the initial time t. This is a simple consequence of the conservation of angular momentum as shown in Appendix A, Exercise 13:

Proposition 2.25 *Let $\gamma : I \to \mathbb{R}^3 \setminus \{0\}$ be the motion of a particle in a radial force field F according to Newton's law $m\gamma'' = F(\gamma)$. Then the angular momentum vector $\mathbf{L}$ is constant, the motion is planar and the area $A(t_0, t_1)$ swept out by the vector γ during a time interval $[t_0, t_1]$ is given by*

$$A(t_0, t_1) = \frac{L}{2}(t_1 - t_0),$$

where L is the length of the angular momentum vector.

As a corollary we obtain *Kepler's third law*:

Corollary 2.26 *The elliptic orbits in Newton's gravitational potential are periodic with period*

$$T = \frac{2\pi}{\sqrt{M}}a^{3/2},$$

where a is the major half-axis of the ellipse.

Proof For a given initial time $t_0 \in \mathbb{R}$ let us denote by $T = T(t_0)$ the smallest positive real number such that $(r, \varphi)|_{t_0+T} = (r, \varphi)_{t_0}$, where $t \mapsto (r(t), \varphi(t))$ is an elliptic motion of the system. Since $A(t_0, t_0 + T) = \frac{L}{2}T$ is the area of the ellipse, T is clearly independent of t_0, which shows that the motion is periodic. The area of the ellipse is $A = \pi ab$ and, hence,

$$T = \frac{2}{L}A = \frac{2\pi ab}{L} \overset{(2.14)}{=} \frac{2\pi ab}{\sqrt{pM}} = \frac{2\pi a^{3/2}}{\sqrt{M}}.$$

$\square$

Chapter 3
Hamiltonian Mechanics

Abstract We present Hamilton's formulation of classical mechanics. In this formulation, the n second-order equations of motion of an n-dimensional mechanical system are replaced by an equivalent set of $2n$ first-order equations, known as Hamilton's equations. There are problems where it is favorable to work with the $2n$ first-order equations instead of the corresponding n second-order equations. After introducing basic concepts from symplectic geometry, we consider the phase space of a mechanical system as a symplectic manifold. We then discuss the relation between Lagrangian and Hamiltonian systems. We show that, with appropriate assumptions, the Euler–Lagrange equations of a Lagrangian mechanical system are equivalent to Hamilton's equations for a Hamiltonian, which can be obtained from the Lagrangian by a Legendre transformation. In the last part, we consider the linearization of mechanical systems as a way of obtaining approximate solutions in cases where the full non-linear equations of motion are too complicated to solve exactly. This is an important tool for analyzing physically realistic theories as these are often inherently non-linear.

3.1 Symplectic Geometry and Hamiltonian Systems

Definition 3.1 A *symplectic manifold* (M, ω) is a smooth manifold M endowed with a *symplectic form* ω, that is a non-degenerate closed 2-form ω.

Example 3.2 (*Symplectic vector space*) Let ω be a non-degenerate skew-symmetric bilinear form on a finite dimensional real vector space V. Then (V, ω) is called a *(real) symplectic vector space*. Every symplectic vector space is of even dimension and there exists a linear isomorphism $V \to \mathbb{R}^{2n}$, $2n = \dim V$, which maps ω to the *canonical symplectic form*

$$\omega_{\text{can}} = \sum_{i=1}^{n} dx^i \wedge dx^{n+i}, \tag{3.1}$$

where $(x^1, \ldots, x^{2n})$ are the standard coordinates on $\mathbb{R}^{2n}$.

© The Author(s) 2017
V. Cortés and A.S. Haupt, *Mathematical Methods of Classical Physics*,
SpringerBriefs in Physics, DOI 10.1007/978-3-319-56463-0_3

It is a basic result in symplectic geometry, known as Darboux's theorem, see e.g. [1, Theorem 3.2.2], that for every point in a symplectic manifold (M, ω) there exists a local coordinate system $(x^1, \ldots, x^{2n})$ defined in a neighborhood U of that point, such that

$$\omega|_U = \sum_{i=1}^{n} dx^i \wedge dx^{n+i}.$$

So $\omega|_U$ looks like the canonical symplectic form on $\mathbb{R}^{2n}$.

Example 3.3 (Cotangent bundle) Let $\pi : N = T^*M \to M$ be the cotangent bundle of a manifold M. We define a 1-form λ on N by

$$\lambda_\xi(v) := \xi(d\pi(v)),$$

for all $\xi \in N$, $v \in T_\xi N$. The 1-form is called the *Liouville form*. Its differential $\omega = d\lambda$ is a symplectic form on N, which is called the *canonical symplectic form* of the cotangent bundle N. To check that ω is indeed non-degenerate, let us compute the Liouville form in coordinates $(q^1, \ldots, q^n, p_1, \ldots, p_n)$ on $\pi^{-1}(U) = T^*U \subset T^*M$ associated with coordinates $(x^1, \ldots, x^n)$ on some open set $U \subset M$:

$$q^i = x^i \circ \pi, \quad p_i(\xi) = \xi\left(\frac{\partial}{\partial x^i}\right), \quad \xi \in T^*U.$$

Since under the projection $d\pi : TN \to TM$ the vector fields $\partial/\partial q^i$ and $\partial/\partial p_i$ are mapped to $\partial/\partial x^i$ and zero, respectively, at every point $\xi = \sum \xi_j dx^j|_{\pi(\xi)} \in T^*U$ we have

$$\lambda_\xi\left(\frac{\partial}{\partial q^i}\right) = \xi\left(\frac{\partial}{\partial x^i}\right) = \xi_i = p_i(\xi), \quad \lambda_\xi\left(\frac{\partial}{\partial p_i}\right) = 0.$$

This shows that

$$\lambda|_U = \sum p_i dq^i, \quad \omega|_U = \sum dp_i \wedge dq^i,$$

proving that ω is a symplectic form.

Given a smooth function f on a symplectic manifold (M, ω), there is a unique vector field X_f such that

$$df = -\omega(X_f, \cdot).$$

Definition 3.4 The vector field X_f is called the *Hamiltonian vector field* associated with f.

According to Darboux's theorem, we can locally write $\omega = \sum dp_i \wedge dq^i$ for some local coordinate system $(q^1, \ldots, q^n, p_1, \ldots, p_n)$ on M. In such coordinates, which will be called *Darboux coordinates*, we can easily compute

$$X_f = \sum \left(\frac{\partial f}{\partial p_i} \frac{\partial}{\partial q^i} - \frac{\partial f}{\partial q^i} \frac{\partial}{\partial p_i} \right), \tag{3.2}$$

see Appendix A, Exercise 19.

Definition 3.5 A *Hamiltonian system* (M, ω, H) is a symplectic manifold (M, ω) endowed with a function $H \in C^\infty(M)$, called the *Hamiltonian*. An integral curve $\gamma : I \to M$ of the Hamiltonian vector field X_H, defined on an open interval $I \subset \mathbb{R}$, is called a *motion* of the Hamiltonian system. The corresponding system of ordinary differential equations

$$\gamma' = X_H(\gamma) \tag{3.3}$$

is called *Hamilton's equation*. An *integral of motion* of the Hamiltonian system is a function $f \in C^\infty(M)$ which is constant along every motion. The symplectic manifold (M, ω) is called the *phase space* of the Hamiltonian system.

Proposition 3.6 *Let (M, ω, H) be a Hamiltonian system. Then H is an integral of motion.*

Proof Let $\gamma : I \to M$ be a motion of the system. Then

$$\frac{d}{dt} H(\gamma(t)) = dH\gamma'(t) = -\omega(X_H(\gamma(t)), \gamma'(t)) = -\omega(X_H, X_H)|_{\gamma(t)} = 0.$$

$\square$

Proposition 3.7 *In Darboux coordinates, Hamilton's equation for a curve $\gamma : I \to M$ takes the form*

$$\dot{q}^i(t) = \frac{\partial H}{\partial p_i}(\gamma(t)), \quad \dot{p}_i(t) = -\frac{\partial H}{\partial q^i}(\gamma(t)), \tag{3.4}$$

for all $i = 1, \ldots, n$ (here, $n = \frac{\dim M}{2}$), where $q^i(t) = q^i(\gamma(t))$, $p_i(t) = p_i(\gamma(t))$.

Proof This follows immediately from (3.2) by comparing $\gamma' = \sum \left(\dot{q}^i \frac{\partial}{\partial q^i} + \dot{p}_i \frac{\partial}{\partial p_i} \right)$ with $X_H(\gamma)$.

$\square$

We refer henceforth to Eq. (3.3) as Hamilton's equation and to Eq. (3.4) as Hamilton's equation*s*.

3.2 Relation between Lagrangian and Hamiltonian Systems

In this section, we clarify the relation between Lagrangian and Hamiltonian systems. We will show, under certain assumptions, that the Euler–Lagrange equations of a Lagrangian mechanical system $(M, \mathscr{L})$ are equivalent to Hamilton's equation for a certain Hamiltonian H on T^*M endowed with the canonical symplectic form ω. If this is true, then the Lagrangian mechanical system $(M, \mathscr{L})$ should possess an integral of motion, corresponding to the integral of motion H of the Hamiltonian system (T^*M, ω, H).

3.2.1 Hamiltonian Formulation for the Lagrangian Systems of Example 2.3

In the case of Lagrangians of the form $\mathscr{L}(v) = \frac{1}{2}g(v, v) - V(\pi(v))$, $v \in TM$, considered in Example 2.3, we showed in Proposition 2.10 that the energy $E(v) = \frac{1}{2}g(v, v) + V(\pi(v))$, $v \in TM$, is indeed an integral of motion. Moreover, we have a natural identification of TM with T^*M by means of the pseudo-Riemannian metric:

$$\phi = \phi_g : TM \to T^*M, \quad v \mapsto g(v, \cdot).$$

Proposition 3.8 *Let (M, g) be a pseudo-Riemannian manifold and denote by ϕ : $TM \to T^*M$ the isomorphism of vector bundles induced by g. Let V be a smooth function on M and consider the Lagrangian $\mathscr{L}$ of Example 2.3.*

(i) *Then a smooth curve $\gamma : I \to M$ is a solution of the Euler–Lagrange equations if and only if the curve $\phi \circ \gamma' : I \to T^*M$ is a solution of Hamilton's equation for the Hamiltonian $H = E \circ \phi^{-1}$.*

(ii) *Conversely, if a smooth curve $\tilde{\gamma} : I \to T^*M$ is a motion of the Hamiltonian system (T^*M, ω, H) then the curve $\pi \circ \tilde{\gamma} : I \to M$ is a motion of the Lagrangian system $(M, \mathscr{L})$, where $\pi : T^*M \to M$ denotes the projection. The maps $\gamma \mapsto \phi \circ \gamma'$ and $\tilde{\gamma} \mapsto \pi \circ \tilde{\gamma}$ are inverse to each other when restricted to solutions of the Euler–Lagrange equations and Hamilton's equations, respectively.*

Proof We prove (i). Part (ii) is similar and part of Appendix A, Exercise 22. Let $(x^1, \ldots, x^n)$ be coordinates defined on some open set $U \subset M$. They induce coordinates $(q^1, \ldots, q^n, \hat{q}^1, \ldots, \hat{q}^n)$ on $TU \subset TM$ and $(q^1, \ldots, q^n, p_1, \ldots, p_n)$ on $T^*U \subset T^*M$. In terms of these coordinates ϕ is given by

$$q^i \circ \phi = q^i, \quad p_i \circ \phi = \sum(g_{ij} \circ \pi)\hat{q}^j = \frac{\partial \mathscr{L}}{\partial \hat{q}^i},$$

where $\pi : TM \to M$ denotes the projection. The inverse $\psi : T^*M \to TM$ is given by

$$q^i \circ \psi = q^i, \quad \hat{q}^i \circ \psi = \sum(g^{ij} \circ \pi)p_j,$$

where (g^{ij}) is the matrix inverse to (g_{ij}) and $\pi : T^*M \to M$ is the projection. Therefore we obtain

$$H = E \circ \psi = \frac{1}{2} \sum (g^{ij} \circ \pi) p_i p_j + V \circ \pi, \quad \pi : T^*M \to M,$$

and by Proposition 3.7, Hamilton's equation takes the form

$$\dot{q}^i = \sum (g^{ij}(\gamma)) p_j = \hat{q}^i,$$

$$\dot{p}_i = -\frac{1}{2} \sum \frac{\partial g^{k\ell}}{\partial x^i}(\gamma) p_k p_\ell - \frac{\partial V}{\partial x^i}(\gamma)$$

$$= \frac{1}{2} \sum \frac{\partial g_{k\ell}}{\partial x^i}(\gamma) \hat{q}^k \hat{q}^\ell - \frac{\partial V}{\partial x^i}(\gamma) = \frac{\partial \mathscr{L}}{\partial q^i}(\gamma'),$$

where $\gamma : I \to M$, $q^i(t) = x^i(\gamma(t))$, $\hat{q}^i(t) = \hat{q}^i(\gamma'(t))$ and $p_i(t) = p_i(\phi(\gamma'(t)))$. So we see that $\phi \circ \gamma' : I \to T^*M$ is a motion of the Hamiltonian system if and only if $\gamma' : I \to TM$ satisfies

$$\dot{q}^i = \hat{q}^i, \quad \dot{p}_i = \frac{\partial \mathscr{L}}{\partial q^i}(\gamma'),$$

where $p_i(t) = \frac{\partial \mathscr{L}}{\partial \hat{q}^i}(\gamma'(t))$. Substituting the expression for p_i into the second equation one obtains precisely the Euler–Lagrange equations, whereas the first equation holds for every curve γ. □

Next we will generalize the above constructions to any Lagrangian satisfying an appropriate non-degeneracy assumption analogous to the non-degeneracy of the pseudo-Riemannian metric.

3.2.2 The Legendre Transform

As a first step we will show that for every Lagrangian mechanical system in the sense of Definition 2.1 (that is for which the Lagrangian does not explicitly depend on time) we can define an integral of motion which generalizes the energy defined for Example 2.3. For this we remark that for the Lagrangian of Example 2.3 the energy can be written in the form

$$E = \sum \frac{\partial \mathscr{L}}{\partial \hat{q}^i} \hat{q}^i - \mathscr{L}.$$

We use this formula to define the *energy* E for any Lagrangian $\mathscr{L}$. To show that the definition is coordinate independent it is sufficient to remark that the vector field

$$\xi = \sum \hat{q}^i \frac{\partial}{\partial \hat{q}^i}$$

is coordinate independent. It is in fact the vector field generated by the one-parameter group of dilatations

$$\varphi_t : TM \to TM, \quad v \mapsto e^t v.$$

Its value at $v \in TM$ is simply

$$\xi(v) = \frac{d}{dt}\Big|_{t=0} \varphi_t(v) = v,$$

where on the right-hand side $v \in T_x M$, $x = \pi(v)$, is interpreted as vertical vector by means of the canonical identification $T_v^{\mathrm{ver}} TM = T_x M$. The following result is a generalization of Proposition 2.10.

Proposition 3.9 (Conservation of energy) *Let $(M, \mathscr{L})$ be a Lagrangian mechanical system. Then the energy $E = \xi(\mathscr{L}) - \mathscr{L}$ is an integral of motion.*

Proof It suffices to differentiate $t \mapsto E(\gamma'(t))$ along a motion $\gamma : I \to M$. We obtain

$$\frac{d}{dt} E(\gamma') = \frac{d}{dt}\left(\sum \frac{\partial \mathscr{L}}{\partial \hat{q}^i} \hat{q}^i - \mathscr{L}\right)$$

$$= \sum \left(\left(\frac{d}{dt}\frac{\partial \mathscr{L}}{\partial \hat{q}^i}\right) \hat{q}^i + \frac{\partial \mathscr{L}}{\partial \hat{q}^i}\dot{\hat{q}}^i - \frac{\partial \mathscr{L}}{\partial q^i}\dot{q}^i - \frac{\partial \mathscr{L}}{\partial \hat{q}^i}\dot{\hat{q}}^i\right),$$

which vanishes by the Euler–Lagrange equations. $\qquad\qquad\qquad\qquad\qquad\square$

Next we generalize the isomorphism $\phi = \phi_g : TM \to T^*M$ defined by a pseudo-Riemannian metric g. As we have shown, it can be written in terms of the Lagrangian as

$$q^i \circ \phi = q^i, \quad p_i \circ \phi = \frac{\partial \mathscr{L}}{\partial \hat{q}^i}.$$

We claim that these formulas define a smooth map $\phi = \phi_{\mathscr{L}} : TM \to T^*M$ for any Lagrangian $\mathscr{L}$. To see this let us define $\phi_{\mathscr{L}}$ in a coordinate independent way. For $v \in T_x M$, $x \in M$, we define

$$\phi_{\mathscr{L}}(v) = d\left(\mathscr{L}|_{T_x M}\right)|_v \in (T_x M)^* = T_x^* M. \tag{3.5}$$

What we obtain is a smooth map that maps any vector $v \in TM$ to a covector $\phi_{\mathscr{L}}(v) \in T_x^* M$ at the point $x = \pi(v)$. In other words, $\phi_{\mathscr{L}}$ is a one-form along the projection map $\pi : TM \to M$, or, equivalently, a section of the vector bundle $\pi^* T^* M$ over TM. Notice that as a map from TM to T^*M it is fiber-preserving but (contrary to ϕ_g) in general not linear on fibers. Also, in general, it is not even a local diffeomorphism. In local coordinates we have

$$\phi_{\mathscr{L}}(v) = \sum \frac{\partial \mathscr{L}(v)}{\partial \hat{q}^i} dx^i|_x.$$

Definition 3.10 Let $(M, \mathscr{L})$ be a Lagrangian mechanical system, $n = \dim M$. Then $\mathscr{L}$ is called *non-degenerate* if $\phi_{\mathscr{L}}$ is of maximal rank, that is if $d\phi_{\mathscr{L}}$ has rank $2n$ everywhere. The Lagrangian is called *nice* if it is non-degenerate and $\phi_{\mathscr{L}}$ is a bijection.

Proposition 3.11 *Let $(M, \mathscr{L})$ be a Lagrangian mechanical system. Then the following conditions are equivalent:*

1. *$\mathscr{L}$ is non-degenerate.*
2. *$\phi_{\mathscr{L}} : TM \to T^*M$ is a local diffeomorphism.*
3. *For all $x \in M$, $\phi_{\mathscr{L}}|_{T_xM} : T_xM \to T_x^*M$ is of maximal rank.*
4. *For all $v \in TM$, there exists a coordinate system (x^i) defined on an open neighborhood U of $\pi(v)$ such that the matrix $\left(\frac{\partial^2 \mathscr{L}(v)}{\partial \hat{q}^i \partial \hat{q}^j} \right)$ is invertible, where $(q^1, \ldots, q^n, \hat{q}^1, \ldots, \hat{q}^n)$ are the corresponding coordinates on TU.*
5. *For all $v \in TM$ and every coordinate system (x^i) defined on an open neighborhood U of $\pi(v)$, the matrix $\left(\frac{\partial^2 \mathscr{L}(v)}{\partial \hat{q}^i \partial \hat{q}^j} \right)$ is invertible.*

Proof See Appendix A, Exercise 20. $\qquad\square$

Proposition 3.12 *Let $(M, \mathscr{L})$ be a Lagrangian mechanical system. Then the following conditions are equivalent:*

1. *$\mathscr{L}$ is nice.*
2. *$\phi_{\mathscr{L}} : TM \to T^*M$ is a diffeomorphism.*
3. *$\mathscr{L}$ is non-degenerate and for all $x \in M$, $\phi_{\mathscr{L}}|_{T_xM} : T_xM \to T_x^*M$ is a bijection.*
4. *For all $x \in M$, $\phi_{\mathscr{L}}|_{T_xM} : T_xM \to T_x^*M$ is a diffeomorphism.*

Proof See Appendix A, Exercise 21. $\qquad\square$

We can now generalize Proposition 3.8 to the case of non-degenerate Lagrangians. For simplicity we assume that $\mathscr{L}$ is nice such that $\phi_{\mathscr{L}} : TM \to T^*M$ is not only a local but a global diffeomorphism. The general (local) result for non-degenerate Lagrangians is left as an exercise (see Appendix A, Exercise 23).

Theorem 3.13 *Let $(M, \mathscr{L})$ be a Lagrangian mechanical system with nice Lagrangian. Then $\phi = \phi_{\mathscr{L}} : TM \to T^*M$ is a diffeomorphism and the following hold:*

(i) *A smooth curve $\gamma : I \to M$ is a solution of the Euler–Lagrange equations if and only if the curve $\phi \circ \gamma' : I \to T^*M$ is a solution of Hamilton's equation for the Hamiltonian $H = E \circ \phi^{-1}$, where E is the energy, defined in Proposition 3.9.*

(ii) *Conversely, if a curve $\tilde{\gamma} : I \to T^*M$ is a motion of the Hamiltonian system (T^*M, ω, H), then the curve $\pi \circ \tilde{\gamma} : I \to M$ is a motion of the Lagrangian system $(M, \mathscr{L})$, where $\pi : T^*M \to M$ denotes the projection.*

(iii) *The maps $\gamma \mapsto \phi \circ \gamma'$ and $\tilde{\gamma} \mapsto \pi \circ \tilde{\gamma}$ are inverse to each other when restricted to solutions of the Euler–Lagrange equations and Hamilton's equations, respectively.*

Proof Let us first remark that the function H on T^*M gives rise to a smooth map $\psi = \psi_H : T^*M \to TM$. For $\alpha \in T_x^*M$, $x \in M$, we define

$$\psi_H(\alpha) = d\left(H|_{T_x^*M}\right)|_\alpha \in (T_x^*M)^* = T_xM. \tag{3.6}$$

In local coordinates (q^i, p_i) on $T^*U \subset T^*M$ and $(q^i, \hat{q}^i)$ on $TU \subset TM$ associated with local coordinates (x^i) on $U \subset M$, it is given by

$$q^i \circ \psi = q^i, \quad \hat{q}^i \circ \psi = \frac{\partial H}{\partial p_i}.$$

We claim that ϕ and ψ are inverse to each other. In particular, ψ is a diffeomorphism. Since we already know that ϕ is bijective, we only need to check that $\psi \circ \phi = \mathrm{Id}_{TM}$. Obviously $q^i \circ \psi \circ \phi = q^i$. Thus it is sufficient to check that $\hat{q}^i \circ \psi \circ \phi = \hat{q}^i$. Let us denote by (a^{ij}) the $n \times n$-matrix inverse to the matrix (a_{ij}) with matrix coefficients $a_{ij} := \frac{\partial^2 \mathscr{L}}{\partial \hat{q}^i \partial \hat{q}^j}$. Then we have

$$\frac{\partial H}{\partial p_i} = dE \circ d\left(\phi^{-1}\right)\frac{\partial}{\partial p_i} = dE \circ (d\phi)^{-1}\frac{\partial}{\partial p_i}$$

$$= \left(dE \sum a^{ij}\frac{\partial}{\partial \hat{q}^j}\right) \circ \phi^{-1} = \left(\sum a^{ij}\frac{\partial E}{\partial \hat{q}^j}\right) \circ \phi^{-1}$$

and

$$\frac{\partial E}{\partial \hat{q}^j} = \frac{\partial}{\partial \hat{q}^j}\left(\sum \hat{q}^k \frac{\partial \mathscr{L}}{\partial \hat{q}^k} - \mathscr{L}\right) = \sum \hat{q}^k \frac{\partial^2 \mathscr{L}}{\partial \hat{q}^j \partial \hat{q}^k} = \sum \hat{q}^k a_{jk}.$$

Substituting the second equation into the first equation, we obtain

$$\frac{\partial H}{\partial p_i} = \hat{q}^i \circ \phi^{-1} \tag{3.7}$$

and, hence,

$$\hat{q}^i \circ \psi \circ \phi = \frac{\partial H}{\partial p_i} \circ \phi = \hat{q}^i.$$

This proves that $\psi = \phi^{-1}$ and, in particular, that $p_i \circ \phi \circ \psi = p_i$, that is

$$\frac{\partial \mathscr{L}}{\partial \hat{q}^i} \circ \psi = p_i. \tag{3.8}$$

To relate Hamilton's equation to the Euler–Lagrange equations, we compute with the help of (3.7) and (3.8):

$$\frac{\partial H}{\partial q^i} = \frac{\partial}{\partial q^i}(E \circ \psi) = \frac{\partial}{\partial q^i}\left(\left(\sum \hat{q}^j \frac{\partial \mathscr{L}}{\partial \hat{q}^j} - \mathscr{L}\right) \circ \psi\right) = \frac{\partial}{\partial q^i}\left(\sum \frac{\partial H}{\partial p_j} p_j - \mathscr{L} \circ \psi\right)$$

$$= \sum \frac{\partial^2 H}{\partial q^i \partial p_j} p_j - \frac{\partial}{\partial q^i}(\mathscr{L} \circ \psi)$$

$$= \sum_j \frac{\partial^2 H}{\partial q^i \partial p_j} p_j - \frac{\partial \mathscr{L}}{\partial q^i} \circ \psi - \sum_j \underbrace{\left(\frac{\partial \mathscr{L}}{\partial \hat{q}^j} \circ \psi\right)}_{=p_j} \underbrace{\frac{\partial(\hat{q}^j \circ \psi)}{\partial q^i}}_{=\frac{\partial^2 H}{\partial q^i \partial p_j}} = -\frac{\partial \mathscr{L}}{\partial q^i} \circ \psi. \qquad (3.9)$$

Now we see from (3.7) and (3.9) that Hamilton's equations for a curve $\tilde{\gamma} : I \to T^*M$ in canonical coordinates (q^i, p_i) take the form

$$\frac{d}{dt}q^i(\tilde{\gamma}(t)) = \frac{\partial H}{\partial p_i}(\tilde{\gamma}(t)) = \hat{q}^i(\psi(\tilde{\gamma}(t))),$$

$$\frac{d}{dt}p_i(\tilde{\gamma}(t)) = -\frac{\partial H}{\partial q^i}(\tilde{\gamma}(t)) = \frac{\partial \mathscr{L}}{\partial q^i}(\psi(\tilde{\gamma}(t))).$$

Notice that the first equation is satisfied if and only if the curve $\psi \circ \tilde{\gamma} : I \to TM$ is the velocity vector field γ' of the curve $\gamma := \pi \circ \tilde{\gamma} : I \to M$. Using $\psi \circ \tilde{\gamma} = \gamma'$, the second equation can be written in the form

$$\frac{d}{dt}(p_i \circ \phi)(\gamma'(t)) = \frac{\partial \mathscr{L}}{\partial q^i}(\gamma'(t)).$$

In view of (3.8) this is equivalent to the Euler–Lagrange equations for the curve γ. So we have proven (ii). This also proves (i) by considering the curve $\tilde{\gamma} = \phi \circ \gamma'$, which projects onto γ.

In order to prove (iii), let us first remark that for every smooth curve γ in M we have $\pi \circ \phi \circ \gamma' = \gamma$, simply because ϕ maps T_xM to T_x^*M for all $x \in M$. Now let $\tilde{\gamma}$ be a solution of Hamilton's equation. Then, as shown above, $\psi \circ \tilde{\gamma} = \gamma'$ is the velocity vector of $\gamma = \pi \circ \tilde{\gamma}$ and, hence, $\phi \circ \gamma' = \tilde{\gamma}$. This proves (iii). □

Let us summarize for completeness some interesting facts, which we have established in the course of the proof.

Proposition 3.14 *Let* $(M, \mathscr{L})$ *be a Lagrangian mechanical system with nice Lagrangian. Then the inverse of the diffeomorphism* $\phi = \phi_{\mathscr{L}} : TM \to T^*M$ *is given by the map* $\psi = \psi_H : T^*M \to TM$ *defined by* $H = E \circ \phi^{-1}$ *in Eq. (3.6). Under these diffeomorphisms the following functions on* TM *and* T^*M *are mapped to each other:*

TM	q^i	$\hat{q}^i$	$\partial \mathscr{L}/\partial \hat{q}^i$	E
T^*M	q^i	$\partial H/\partial p_i$	p_i	H

Next we will explain the relation of the previous constructions with the notion of Legendre transform of a smooth function $f : V \to \mathbb{R}$ on a finite-dimensional real vector space V. To define the Legendre transform we consider the smooth map

$$\phi_f : V \to V^*, \quad x \mapsto df_x. \tag{3.10}$$

For simplicity we will assume that ϕ_f is a diffeomorphism. (More generally, we could consider the case when ϕ_f is only locally a diffeomorphism.) Then we can define a new function $\tilde{f} : V^* \to \mathbb{R}$ by

$$\tilde{f} := (\xi(f) - f) \circ \phi_f^{-1}, \tag{3.11}$$

where ξ is the position vector field in V, that is $\xi_x = x$ for all $x \in V$. Evaluating this function at $y = \phi_f(x)$, we obtain

$$\tilde{f}(y) = \xi_x(f) - f(x) = df_x x - f(x) = \langle y, x \rangle - f(x),$$

where $\langle y, x \rangle = yx = y(x)$ is the duality pairing. This shows that (3.11) can be equivalently written as

$$\tilde{f}(y) = (\langle y, x \rangle - f(x))|_{x=\phi_f^{-1}(y)}. \tag{3.12}$$

Definition 3.15 The function $\tilde{f} : V^* \to \mathbb{R}$ is called the *Legendre transform* of $f : V \to \mathbb{R}$.

Proposition 3.16 *Let f be a smooth function on a finite-dimensional real vector space V such that $\phi_f : V \to V^*$ is a diffeomorphism and consider its Legendre transform $\tilde{f} \in C^\infty(V^*)$. Then $\phi_{\tilde{f}} : V^* \to V$ is a diffeomorphism and the Legendre transform of $\tilde{f}$ is f.*

Proof See Appendix A, Exercise 24.   □

Notice that, with the above notations, for every Lagrangian $\mathscr{L} \in C^\infty(TM)$ the restriction of $\phi_{\mathscr{L}} : TM \to T^*M$, defined in (3.5), to $T_x M$, $x \in M$, is given by

$$\phi_{\mathscr{L}}|_{T_x M} = \phi_{\mathscr{L}_x}, \quad \mathscr{L}_x := \mathscr{L}|_{T_x M} : T_x M \to \mathbb{R}.$$

Similarly, for every function $H \in C^\infty(T^*M)$

$$\psi_H|_{T_x^* M} = \phi_{H_x}, \quad H_x := H|_{T_x^* M} : T_x^* M \to \mathbb{R}.$$

In view of this relation, we will now unify the notation and define $\phi_H := \psi_H$.

Proposition 3.17 *Let $\mathscr{L} \in C^\infty(TM)$ be a nice Lagrangian and $H \in C^\infty(T^*M)$ the corresponding Hamiltonian. Then H is the fiber-wise Legendre transform of $\mathscr{L}$ and vice versa, that is H_x is the Legendre transform of $\mathscr{L}_x$ and $\mathscr{L}_x$ is the Legendre transform of H_x for all $x \in M$.*

Proof By comparing the definition of the energy $E = \xi(\mathscr{L}) - \mathscr{L}$ with (3.11) we see that $H_x = E_x \circ \phi_{\mathscr{L}_x}^{-1} = \tilde{\mathscr{L}}_x$ and this implies that $\tilde{H}_x = \mathscr{L}_x$ by Proposition 3.16.   □

3.3 Linearization and Stability

The Euler–Lagrange equations and Hamilton's equations are typically non-linear. We refer to Example 2.7 for an illustration of this observation. Non-linearity is also a key feature in many interesting physical applications. Examples from classical mechanics are the aerodynamic drag, where the drag force is proportional to the square of the velocity, and the Navier-Stokes equations describing the motion of viscous fluids. Important examples from modern physics are Einstein's theory of general relativity (see Sect. 5.4.3) as well as the Standard Model of particle physics.

However, non-linear equations are notoriously difficult to solve. A way out is to consider a linear approximation, instead. This often yields valuable insights into the true behavior of the underlying non-linear problem and serves as a starting point for more thorough studies (such as perturbation theory).

This section is based on Ref. [2, Chap. 5]. In this section we denote by $I \subset \mathbb{R}$ an open interval.

Definition 3.18 A point $x_0 \in \mathbb{R}^n$ is called an *equilibrium position* of the system of ordinary differential equations

$$\frac{dx}{dt} = f(x), \quad x : I \to \mathbb{R}^n, \tag{3.13}$$

if the constant curve $x(t) = x_0$ is a solution.

Notice that the equilibrium positions of the system (3.13) are precisely the zeroes of f.

For the rest of this section, we consider as the prime example a classical mechanical system on $M = \mathbb{R}^n$ with canonical local coordinates $(q^1, \ldots, q^n, \hat{q}^1, \ldots, \hat{q}^n)$ on TM and Lagrangian

$$\mathcal{L} = \frac{1}{2} \sum a_{ij}(q) \hat{q}^i \hat{q}^j - V(q). \tag{3.14}$$

The functions $a_{ij}(q)$ are chosen such that $E_{\text{kin}} = \frac{1}{2} \sum a_{ij}(q) \hat{q}^i \hat{q}^j > 0$ for all $\hat{q} \neq 0$. The motion is governed by the Euler–Lagrange equations

$$\frac{d}{dt} \frac{\partial \mathcal{L}}{\partial \hat{q}^i} - \frac{\partial \mathcal{L}}{\partial q^i} = 0, \quad i = 1, \ldots, n. \tag{3.15}$$

Proposition 3.19 *The point* $q = q_0$, $\hat{q} = \hat{q}_0$ *is an equilibrium position of the Lagrangian mechanical system* (3.14) *if and only if* $\hat{q}_0 = 0$ *and* q_0 *is a critical point of* $V(q)$, *that is*

$$\left. \frac{\partial V}{\partial q^i} \right|_{q=q_0} = 0 \quad \forall i = 1, \ldots, n. \tag{3.16}$$

Proof From Proposition 3.7 we know that the Euler–Lagrange equations (3.15) can be transformed into a system of $2n$ first-order equations of the form

$$\dot{q}^i = \sum a^{ij} p_j = \hat{q}^i \,, \qquad \dot{p}_i = -\frac{1}{2} \sum \frac{\partial a^{k\ell}}{\partial q^i} p_k p_\ell - \frac{\partial V}{\partial q^i} \,,$$

where $(q^1, \ldots, q^n, p_1, \ldots, p_n)$ are local coordinates on T^*M and (a^{ij}) denotes the inverse matrix of (a_{ij}). At an equilibrium position, we have $\dot{q} = \dot{p} = 0$ and hence from the first equation $\hat{q} = 0$, $p = 0$. With $p = 0$ the first term in the second equation vanishes and hence we conclude that $q = q_0$ is an equilibrium position if (3.16) holds and only in that case. □

Returning to the general case, $\frac{dx}{dt} = f(x)$, we may Taylor-expand $f(x)$ close to an equilibrium position x_0. For convenience, we may assume without loss of generality $x_0 = 0$ (by a translation of the coordinate system). The Taylor expansion close to $x_0 = 0$ then becomes

$$f(y) = Ay + \mathcal{O}(|y|^2) \,, \tag{3.17}$$

where $A = \frac{\partial f}{\partial x}|_{x=0}$.

Definition 3.20 The passage from the system

$$\frac{dx}{dt} = f(x) \,, \qquad x : I \to \mathbb{R}^n \,,$$

to the linear system

$$\frac{dy}{dt} = Ay \,, \qquad y : I \to T_0\mathbb{R}^n \cong \mathbb{R}^n \,,$$

is called *linearization* around the equilibrium position $x_0 = 0$.

The linearized system can be easily solved by

$$y(t) = e^{At} y(0) \,,$$

where $e^{At} = \mathbb{1}_n + At + \frac{1}{2}A^2 t^2 + \ldots$ is the matrix exponential series. For small enough y, the higher-order corrections $\mathcal{O}(|y|^2)$ in (3.17) are small compared to y itself. Thus, for a long time, the solutions $y(t)$ of the linear system and $x(t)$ of the full system remain close to each other, provided that the initial conditions $y(0) = x(0)$ are chosen sufficiently close to the equilibrium position x_0. More precisely, for a given time $T > 0$ and $\varepsilon > 0$, there exists a $\delta > 0$ such that for any $x_0' \in \mathbb{R}^n$ with $|x_0' - x_0| < \delta$ the solution $x(t)$ of (3.13) with initial condition $x(0) = x_0'$ exists (at least) for all $t \in [0, T]$ and compares to the solution $y : \mathbb{R} \to \mathbb{R}^n$ of the linearized system with $y(0) = x_0'$ by

$$\max_{[0,T]} |x - y| < \varepsilon \,.$$

Consider the Lagrangian mechanical system (3.14) near the equilibrium position $q = q_0$ and choose coordinates such that $q_0 = 0$.

Proposition 3.21 *The linearization of (3.14) near $q = q_0$ is given by*

$$\mathscr{L}_2 = \frac{1}{2}\sum a_{ij}(0)\hat{q}^i\hat{q}^j - \frac{1}{2}\sum \frac{\partial^2 V}{\partial q^i \partial q^j}\Big|_{q=0} q^i q^j . \tag{3.18}$$

This is also known as quadratic approximation.

Proof The Hamiltonian corresponding to (3.14) is given by

$$H = \frac{1}{2}\sum a^{ij}(q)p_i p_j + V(q) .$$

Hamilton's equations can be written as

$$\dot{p}_i = -\frac{\partial H}{\partial q^i} =: f_i(p, q) , \qquad \dot{q}^i = \frac{\partial H}{\partial p_i} =: g^i(p, q) ,$$

The linearization of this system is obtained by Taylor expanding f and g around $q = p = 0$ keeping only terms that are at most linear in p and q:

$$\dot{p}_i = \sum \frac{\partial f_i}{\partial q^j}\Big| q^j + \sum \frac{\partial f_i}{\partial p_j}\Big| p_j = -\sum \frac{\partial^2 H}{\partial q^i \partial q^j}\Big| q^j - \sum \frac{\partial^2 H}{\partial q^i \partial p_j}\Big| p_j$$

$$= -\sum \frac{\partial^2 V}{\partial q^i \partial q^j}\Big| q^j - \frac{1}{2}\sum \left(\frac{\partial^2 a^{k\ell}}{\partial q^i \partial q^j} p_k p_\ell\right)\Big| q^j - \sum \left(\frac{\partial a^{jk}}{\partial q^i} p_k\right)\Big| p_j$$

$$= -\sum \frac{\partial^2 V}{\partial q^i \partial q^j}\Big| q^j ,$$

$$\dot{q}^i = \sum \frac{\partial g^i}{\partial q^j}\Big| q^j + \sum \frac{\partial g^i}{\partial p_j}\Big| p_j = \sum \frac{\partial^2 H}{\partial p_i \partial q^j}\Big| q^j + \sum \frac{\partial^2 H}{\partial p_i \partial p_j}\Big| p_j$$

$$= \sum \left(\frac{\partial a^{ik}}{\partial q^j} p_k\right)\Big| q^j + \sum a^{ij}(0)p_j = \sum a^{ij}(0)p_j .$$

In the above equations a vertical line is used as a shorthand symbol to denote the evaluation of the preceding expression at $q = p = 0$. Differentiating and combining the two equations yields

$$\ddot{q}^i = \sum a^{ij}(0)\dot{p}_j = -\sum a^{ij}(0)\frac{\partial^2 V}{\partial q^j \partial q^k}\Big|_{q=0} q^k .$$

These are precisely the Euler–Lagrange equations obtained from $\mathscr{L}_2$. $\qquad\square$

Example 3.22 Consider the case $n = 1$:

$$\mathcal{L} = \frac{1}{2}a(q)\dot{q}^2 - V(q) .$$

Let $q(t) = q_0$ be an equilibrium position, that is

$$\left.\frac{\partial V}{\partial q}\right|_{q=q_0} = 0 .$$

Assuming without loss of generality $q_0 = 0$, the linearized Lagrangian becomes

$$\mathcal{L}_2 = \frac{1}{2}\alpha\dot{q}^2 - \frac{1}{2}\beta q^2 ,$$

where $\alpha := a(0)$ and $\beta := \left.\frac{\partial^2 V}{\partial q^2}\right|_{q=q_0}$. Note that $\alpha > 0$ by assumption (cf. below Eq. (3.14)). The Euler–Lagrange equation corresponding to $\mathcal{L}_2$ is given by

$$\ddot{q} = -\omega_0^2 q , \qquad \omega_0^2 := \frac{\beta}{\alpha} . \tag{3.19}$$

The motion crucially depends on the sign of β or, in other words, on whether the potential $V(q)$ attains a local minimum or maximum at the equilibrium position q_0. Indeed, we find as solution of (3.19) with integration constants $c_1, c_2 \in \mathbb{R}$,

$$q(t) = \begin{cases} c_1 \cos(\omega_0 t) + c_2 \sin(\omega_0 t) , & \text{for } \beta > 0 \text{ (``small oscillations'') ,} \\ c_1 \cosh(|\omega_0|t) + c_2 \sinh(|\omega_0|t) , & \text{for } \beta < 0 \text{ (``runaway behavior'') ,} \\ c_1 t + c_2 , & \text{for } \beta = 0 \text{ (``uniform motion'') .} \end{cases}$$

This observation leads to the notion of *stability*.

Now, we consider the case of higher-dimensional motion on $\mathbb{R}^n$, $n > 1$.

Definition 3.23 Consider a Lagrangian mechanical system of the form (3.14) with equilibrium position $q = q_0$ and set

$$(\Omega)_{ij} := \left.\frac{\partial^2 V}{\partial q^i \partial q^j}\right|_{q=q_0} , \qquad i, j = 1, \ldots, n .$$

The equilibrium position $q = q_0$ is called

  (i) *unstable*, if Ω is negative definite,
 (ii) *stable*, if Ω is positive definite,
(iii) *degenerate*, if Ω is degenerate,
 (iv) a *saddle point*, if Ω is nondegenerate but indefinite.

This can be understood as follows. Condition (i) implies that the potential $V(q)$ attains a local maximum at the critical point q_0. Hence, any small displacement away from q_0 causes the system to deviate further from it, since the net-force is directed away from q_0. An example of such a situation is depicted in the following figure:

In case (ii), $V(q)$ attains a local minimum at $q = q_0$. For small perturbations, the system stays close to q_0 for all times, since the forces are "restoring forces." This can be exemplified as follows:

A typical example of case (iii) is a region where the potential $V(q)$ is flat:

In that case, higher-order derivatives are needed to decide stability. Finally, condition (iv) can be understood as a situation where some directions are stable and some are unstable.

For the rest of this section, to streamline the presentation we only consider stable equilibrium positions. The general case can be analyzed similarly. For one-dimensional motion we saw in Example 3.22 that the linearized motion around a stable equilibrium position is oscillatory in nature,

$$q(t) = c_1 \cos(\omega_0 t) + c_2 \sin(\omega_0 t) .$$

Here, $\omega_0 = \sqrt{\beta/\alpha}$ is the *frequency*, and $\tau_0 = 2\pi/\omega_0$ the *period* of the oscillation. This leads us to the next concept, namely *small oscillations*.

Definition 3.24 Motions in a linearized system $\mathscr{L}_2$, as defined in (3.18), are called *small oscillations* near an equilibrium position $q = q_0$.

Recall that the linearized Lagrangian is given by

$$\mathscr{L}_2 = \frac{1}{2} \sum \alpha_{ij} \dot{q}^i \dot{q}^j - \frac{1}{2} \sum \Omega_{ij} q^i q^j .$$

where $\alpha_{ij} := a_{ij}(0)$ and $\Omega_{ij} := \left. \frac{\partial^2 V}{\partial q^i \partial q^j} \right|_{q=0}$. Note that (α_{ij}) and $\Omega := (\Omega_{ij})$ are symmetric, real matrices and (α_{ij}) is positive definite. It is useful to work in coordinates where $\alpha_{ij} = \delta_{ij}$, where δ_{ij} is the *Kronecker delta*.

The equations of motion following from $\mathscr{L}_2$ are a set of a priori coupled linear ordinary differential equations,

$$\ddot{q}^i = -\sum \delta^{ij} \Omega_{jk} q^k .$$

Here, the Kronecker delta δ^{ij} (note that $\delta^{ij} = \delta_{ij}$) is written explicitly in order to make it manifest that the index positions on both sides of the equation match. By a suitable choice of coordinates, we can *de-couple* the ordinary differential equations. That is, we *diagonalize* the real symmetric matrix Ω using an orthogonal transformation $M \in O(n)$, so that

$$M^\mathsf{T} \Omega M = \mathrm{diag}(\lambda_1, \ldots, \lambda_n) ,$$

where $\lambda_1, \ldots, \lambda_n \in \mathbb{R}$ are the eigenvalues of Ω and $(\cdot)^\mathsf{T}$ denotes matrix transposition. Now, we define new coordinates Q on $\mathbb{R}^n$ as

$$Q = M^\mathsf{T} q .$$

In terms of the new coordinates, the equations of motion become

$$\ddot{Q}^i = -\lambda_i Q^i \qquad \text{(no sum over } i) .$$

The corresponding Lagrangian is given by

$$\mathscr{L}_2 = \frac{1}{2} \sum_i \left(\dot{Q}^i \right)^2 - \frac{1}{2} \sum_i \lambda_i \left(Q^i \right)^2 ,$$

up to an additive constant. The de-coupled system of equations can be solved straightforwardly. One finds n independent harmonic oscillators of the form

$$Q^i(t) = c_1^i \cos \left(\sqrt{\lambda_i}\, t \right) + c_2^i \sin \left(\sqrt{\lambda_i}\, t \right) \qquad \text{(no sum over } i) ,$$

with real integration constants $(c_{1,2}^i)_{i=1,\ldots,n}$. Recall that the stability criterion (see Definition 3.23) guarantees $\lambda_i > 0$ for all i. Otherwise the linearized motion will in general not be periodic, but solutions can be described in a similar way using hyperbolic and linear functions for the cases $\lambda_i < 0$ and $\lambda_i = 0$, respectively.

What we have just learned is that a system performing small oscillations decomposes into a direct product of n one-dimensional systems performing small oscillations. In particular, the system can perform small oscillations of the form

$$q(t) = [c_1 \cos(\omega t) + c_2 \sin(\omega t)] \xi , \tag{3.20}$$

where $\omega = \sqrt{\lambda}$ and ξ is an eigenvector of Ω corresponding to λ, that is

$$\Omega \xi = \lambda \xi .$$

This oscillation can be regarded as a product of $Q^i = c_1^i \cos(\omega_i t) + c_2^i \sin(\omega_i t)$ and $Q^j = 0$, $j \neq i$, for some $i \in \{1, \ldots, n\}$. The two-parameter family of solutions (3.20) is called a *characteristic oscillation* or *eigen-oscillation* or *normal mode* and $\omega = \sqrt{\lambda}$ is called *characteristic frequency* or *eigen-frequency* (or sometimes also *resonance frequency*). Sometimes also an element of that family is called a characteristic oscillation. The vector ξ is called the *eigenvector* corresponding to the characteristic oscillation and a system of characteristic oscillation is called *independent* if the corresponding eigenvectors are linearly independent.

The above results are summarized in the following theorem.

Theorem 3.25 *The linearized Lagrangian mechanical system of the form* (3.18) *near a stable equilibrium position* $q = q_0$ *performs small oscillations given by a sum of characteristic oscillations. The system has n independent characteristic oscillations and the characteristic frequencies are the square roots of the eigenvalues of the Hessian matrix of the potential* $V(q)$ *at* q_0 *(assuming that* $\alpha_{ij} = \delta_{ij}$, *without loss of generality).*

Note that a *sum* of characteristic oscillations is generally *not periodic*.[1]

The linearized Lagrangian mechanical system $\mathcal{L}_2$ can now be solved in the following way:

(i) Find the complex characteristic oscillations of the form $q(t) = e^{i\omega t}\xi$ by substituting into the equations of motion $\ddot{q} = -\Omega q$. This yields a characteristic equation, $\Omega\xi = \omega^2\xi$. Solving this equation produces n pairwise orthogonal eigenvectors ξ_k with corresponding real eigenvalues $\lambda_k = \omega_k^2$.

(ii) The general real-valued solution is a linear combination of (i). That is,

$$q(t) = \text{Re} \sum_{k=1}^{n} c_k e^{i(\omega_k t + \delta_k)} \xi_k \ ,$$

with c_k and δ_k real parameters.

Remark This result is valid irrespective of the multiplicities of the eigenvalues λ_k.

Example 3.26 (*See* [2, Chap. 5] *for this and further examples*) Consider two identical mathematical pendula of unit mass connected by a weightless spring as depicted in the following figure:

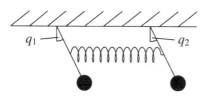

---

[1]For example, consider a case with two eigenvalues $\lambda_1 = 1$, $\lambda_2 = 2$, and the solution $q(t) = \sin(t)\xi_1 + \sin(\sqrt{2}\,t)\xi_2$.

For small oscillations, we have $E_{kin} = \frac{1}{2}\dot{q}_1^2 + \frac{1}{2}\dot{q}_2^2$ and $E_{pot} = V = \frac{1}{2}q_1^2 + \frac{1}{2}q_2^2 + \frac{k}{2}(q_1 - q_2)^2$, where the last term in V is due to the spring. We choose new diagonalizing coordinates

$$Q_1 = \frac{q_1 + q_2}{\sqrt{2}}, \qquad Q_2 = \frac{q_1 - q_2}{\sqrt{2}}.$$

In terms of the new coordinates, the linearized Lagrangian becomes

$$\mathcal{L}_2 = E_{kin} - E_{pot} = \frac{1}{2}\dot{Q}_1^2 + \frac{1}{2}\dot{Q}_2^2 - \frac{1}{2}\omega_1^2 Q_1^2 - \frac{1}{2}\omega_2^2 Q_2^2,$$

with $\omega_1 = 1$ and $\omega_2 = \sqrt{1 + 2k}$. There are two characteristic oscillations, namely

(a) $Q_2 = 0$, that is $q_1 = q_2$.

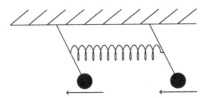

This is the case where both pendula move in phase and the spring has no effect.

(b) $Q_1 = 0$, that is $q_1 = -q_2$.

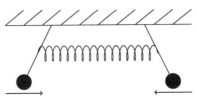

In this case, the pendula move in opposite phase with increased frequency $\omega_2 > 1$ due to the presence of the spring.

Chapter 4
Hamilton–Jacobi Theory

Abstract Besides the Newtonian, Lagrangian, and Hamiltonian formulations of classical mechanics, there is yet a fourth approach, known as Hamilton–Jacobi theory, which is part of Hamiltonian mechanics. This approach is the subject of the present chapter. In Hamilton–Jacobi theory, the central equation capturing the dynamics of the mechanical system is the Hamilton–Jacobi equation which is a first-order, non-linear partial differential equation. Remarkably and contrary to the other formulations of classical mechanics, the entire multi-dimensional dynamics is described by a single equation. Even for relatively simple mechanical systems, the corresponding Hamilton–Jacobi equation can be hard or even impossible to solve analytically. However, its virtue lies in the fact that it offers a useful, alternative way of identifying conserved quantities even in cases where the Hamilton–Jacobi equation itself cannot be solved directly. In addition, Hamilton–Jacobi theory has played an important historical role in the development of quantum mechanics, since the Hamilton–Jacobi equation can be viewed as a precursor to the Schrödinger equation [3, 7, 18].

For simplicity we consider Hamiltonian systems (T^*M, H) of cotangent type, that is for which the phase space is the cotangent bundle T^*M of a (connected) manifold with canonical symplectic form $\omega = \omega_{\text{can}}$. As we have shown in Chap. 3, this includes the systems obtained from Lagrangian mechanical systems with a nice Lagrangian via Legendre transform. By Darboux's theorem every Hamiltonian system can be locally identified with an open subset of a Hamiltonian system of cotangent type.

We will now describe a method for the solution of Hamilton's equation which is based on the following simple idea. Consider a function S on the base manifold M. It gives rise to a 1-form $dS : M \to T^*M$ and for every curve $\gamma : I \to M$ we can consider the curve

$$\tilde{\gamma} : I \to T^*M, \quad t \mapsto dS_{\gamma(t)}.$$

© The Author(s) 2017
V. Cortés and A.S. Haupt, *Mathematical Methods of Classical Physics*,
SpringerBriefs in Physics, DOI 10.1007/978-3-319-56463-0_4

We would like to know when $\tilde{\gamma}$ is an integral curve of the Hamiltonian vector field X_H. Since H is an integral of motion, a necessary condition is that $H \circ \tilde{\gamma}$ is constant. Projecting Hamilton's equation $\tilde{\gamma}' = X_H(\tilde{\gamma})$ to M we obtain

$$\gamma' = d\pi \, X_H(\tilde{\gamma}). \tag{4.1}$$

We will prove in the next theorem that, conversely, this equation (for $n = \dim M$ functions) is sufficient to solve Hamilton's equation (for $2n$ functions) if we assume that H is constant on the section dS of T^*M. Notice that in canonical coordinates (q^i, p_i) associated with local coordinates (x^i) in M the Eq. (4.1) corresponds to the system

$$\dot{q}^i = \frac{\partial H}{\partial p_i}(\tilde{\gamma}), \quad i = 1, \ldots, n,$$

where $q^i(t) = q^i(\tilde{\gamma}(t)) = x^i(\gamma(t))$. This is exactly half (that is, n out of $2n$) of Hamilton's equations.

Theorem 4.1 Let (T^*M, H) be a Hamiltonian system of cotangent type and $S \in C^\infty(M)$. Then the following are equivalent.

1. For all solutions γ of (4.1), the curve $\tilde{\gamma} = dS_\gamma$ is a solution of Hamilton's equation.
2. The Hamiltonian vector field X_H is tangent to the image of $dS : M \to T^*M$.
3. The function $H \circ dS : M \to \mathbb{R}$ is constant.

To prove this theorem we will use the following fundamental lemma.

Lemma 4.2 Let α be a smooth 1-form on M. We consider it as a smooth map

$$\varphi_\alpha : M \to T^*M.$$

Then the following holds:

(i) The pull back of the Liouville form λ under this map is given by

$$\varphi_\alpha^* \lambda = \alpha.$$

(ii) The pull-back of the canonical symplectic form ω is

$$\varphi_\alpha^* \omega = d\alpha.$$

Proof (i) For $v \in TM$, we compute

$$(\varphi_\alpha^* \lambda)(v) = \lambda(d\varphi_\alpha v) = \alpha(d\pi \, d\varphi_\alpha v) = \alpha(v).$$

Part (ii) follows from (i):

$$\varphi_\alpha^* \omega = \varphi_\alpha^* d\lambda = d\varphi_\alpha^* \lambda = d\alpha.$$

□

Definition 4.3 An immersion $\varphi : N \to M$ of an n-dimensional manifold N into a $2n$-dimensional symplectic manifold (M, ω) is called *Lagrangian* if it satisfies $\varphi^* \omega = 0$.

Lemma 4.2 (ii) immediately implies:

Corollary 4.4 *The embedding $\varphi_\alpha : M \to T^*M$ defined by a smooth 1-form α on a manifold M is Lagrangian if and only if α is closed.*

Definition 4.5 Let (V, ω) be a symplectic vector space. A subspace $U \subset V$ is called *Lagrangian* if U coincides with

$$U^\perp = \{v \in V \mid \omega(v, u) = 0 \quad \text{for all} \quad u \in U\}.$$

It is easy to prove the following proposition (see Appendix A, Exercise 25).

Proposition 4.6 *An immersion $\varphi : N \to M$ of an n-dimensional manifold N into a $2n$-dimensional symplectic manifold (M, ω) is Lagrangian if and only if its tangent spaces $d\varphi T_x N \subset T_{\varphi(x)} M$ are Lagrangian for all $x \in N$.*

Proof (of Theorem 4.1) Let us first show that 1. and 2. are equivalent. We first assume 1. and show that $X_H(\varphi_{dS}(x))$ is tangent to the Lagrangian submanifold $N = \varphi_{dS}(M) \subset T^*M$ for all $x \in M$. Let $\gamma : I = (-\varepsilon, \varepsilon) \to M, \varepsilon > 0$, be a solution of (4.1) with initial condition $\gamma(0) = x$. Then the curve $\tilde{\gamma} = \varphi_{dS} \circ \gamma$ is a solution to Hamilton's equation and lies in N. Therefore,

$$X_H(\tilde{\gamma}) = \tilde{\gamma}' \in TN$$

and, in particular, $X_H(\varphi_{dS}(x)) = X_H(\tilde{\gamma}(0)) = \tilde{\gamma}'(0) \in TN$. So we have proven that 1. implies 2. To show the converse, we observe that given a solution $\gamma : I \to M$ of (4.1) and $t \in I$, the vector $\tilde{\gamma}'(t) \in T_{\tilde{\gamma}(t)} N$ is the unique tangent vector in $T_{\tilde{\gamma}(t)} N$ which projects to $\gamma'(t) = d\pi X_H(\tilde{\gamma}(t))$. Since the vector $X_H(\tilde{\gamma}(t)) \in T_{\tilde{\gamma}(t)} T^*M$ projects to the same vector in the base, we see that $\tilde{\gamma}'(t) = X_H(\tilde{\gamma}(t))$ if and only if $X_H(\tilde{\gamma}(t)) \in TN$.

Next we prove the equivalence of 2. and 3. Since $N \subset T^*M$ is Lagrangian, the vector field X_H is tangent to N if and only if at all points $u \in N$ we have

$$X_H(u) \in (T_u N)^\perp \iff dH T_u N = 0 \iff d(H \circ \varphi_{dS})|_{\pi(u)} = 0.$$

□

The partial differential equation

$$H \circ dS = \text{const}$$

of part 3 of Theorem 4.1 is called the *Hamilton–Jacobi equation*. We can summarize part of Theorem 4.1 as follows. If S is a solution to the Hamilton–Jacobi equation then every integral curve of X_H passing through a point of the Lagrangian submanifold $N = \varphi_{dS}(M) \subset T^*M$ is fully contained in N and these integral curves are obtained by solving a system of n ordinary differential equations (rather than $2n$).

Next we will show that given not only one solution, but a smooth family of solutions of the Hamilton–Jacobi equation depending on n parameters we can (at least locally) completely solve Hamilton's equations of motion, provided the family is non-degenerate in a certain sense. The non-degeneracy assumption will ensure that the family cannot be locally reduced to a family depending on fewer parameters.

Definition 4.7 Let (T^*M, H) be a Hamiltonian system of cotangent type and $n = \dim M$. A *family* of solutions of the Hamilton–Jacobi equation is a smooth function

$$S : M \times U \to \mathbb{R}, \quad (x, u) \mapsto S(x, u),$$

where $U \subset \mathbb{R}^n$ is an open subset, such that for all $u \in U$ the function $x \mapsto S^u(x) := S(x, u)$ on M is a solution of the Hamilton–Jacobi equation. It is called *non-degenerate* if the map

$$\Phi_S : M \times U \to T^*M, \quad (x, u) \mapsto dS^u|_x \in T_x^*M \subset T^*M$$

is of maximal rank.

Let $(S^u)_{u \in U}$ be a smooth family of solutions of the Hamilton–Jacobi equation. Then, for all $u \in U$, the image of $dS^u : M \to T^*M$ is a Lagrangian submanifold N_u. The smooth map Φ_S maps each submanifold $M \times \{u\} \subset M \times U$ diffeomorphically to the submanifold $N_u \subset T^*M$. Moreover, it is fiber preserving in the sense that it maps the fibers $\{x\} \times U$, $x \in M$, of the trivial projection $M \times U \to M$ into the fibers T_x^*M of the cotangent bundle. The next proposition is left as an exercise (see Appendix A, Exercise 26).

Proposition 4.8 *Let (T^*M, H) be a Hamiltonian system of cotangent type, $n = \dim M$ and let $S : M \times U \to \mathbb{R}$ be a smooth n-parameter family of solutions of the Hamilton–Jacobi equation. Then the following conditions are equivalent.*

(i) The family is non-degenerate.
(ii) $\Phi_S : M \times U \to T^*M$ is a local diffeomorphism.
(iii) For all $x \in M$,

$$\Phi_S|_{\{x\} \times U} : \{x\} \times U \cong U \to T_x^*M, \quad u \mapsto dS^u|_x,$$

is a local diffeomorphism.

(iv) For all $x \in M$,

$$\Phi_S|_{\{x\} \times U} : \{x\} \times U \cong U \to T_x^* M,$$

is of maximal rank.

(v) For all $(x, u) \in M \times U$ the $n \times n$-matrix

$$\left(\frac{\partial^2 S(x, u)}{\partial x^i \partial u^j} \right)$$

is invertible, where (x^i) are local coordinates in a neighborhood of $x \in M$ and (u^i) are (for instance) standard coordinates in $U \subset \mathbb{R}^n$.

Lemma 4.9 *Let (T^*M, H) be a Hamiltonian system of cotangent type, $n = \dim M$, and $S : M \times U \to \mathbb{R}$ a smooth non-degenerate n-parameter family of solutions of the Hamilton–Jacobi equation. Then the pull back of the canonical symplectic form by the local diffeomorphism $\Phi_S : M \times U \to T^*M$ is given by*

$$\Phi_S^* \omega = \sum \frac{\partial^2 S}{\partial x^i \partial u^j} dx^i \wedge du^j,$$

with respect to any local coordinate system $(x^1, \ldots, x^n)$ on M, where $(u^1, \ldots, u^n)$ are standard coordinates on $U \subset \mathbb{R}^n$.

Proof Let us denote by (q^i, p_i) the local coordinates on T^*M associated with the local coordinates (x^i) on M. Then

$$q^i \circ \Phi_S = x^i, \quad p_i \circ \Phi_S = \frac{\partial S}{\partial x^i}$$

and, hence,

$$\Phi_S^* \lambda = \sum \frac{\partial S}{\partial x^i} dx^i \implies \Phi_S^* \omega = d\Phi_S^* \lambda = \sum \frac{\partial^2 S}{\partial x^i \partial u^j} dx^i \wedge du^j.$$

$\square$

Theorem 4.10 *Let (T^*M, H) be a Hamiltonian system of cotangent type, where M is connected, $n = \dim M$ and $S : M \times U \to \mathbb{R}$ a smooth non-degenerate n-parameter family of solutions of the Hamilton–Jacobi equation. Then the motions of the Hamiltonian system (T^*M, H) contained in the open subset $\Phi_S(M \times U) \subset T^*M$ are of the form $\Phi_S \circ \gamma$, where $\gamma : I \to M \times U$ is a motion of the Hamiltonian system*

$$(M \times U, \ \tilde{\omega} := \phi_S^* \omega, \ \tilde{H} := H \circ \Phi_S).$$

The equations of motion of the latter system can be solved by passing to local coordinates of the form

$$\left(\tilde{q}^i := u^i, \ \tilde{p}_i := -\frac{\partial S}{\partial u^i} \right).$$

In such coordinates we have

$$\tilde{\omega} = \sum d\tilde{q}^i \wedge d\tilde{p}_i, \quad \frac{\partial \tilde{H}}{\partial \tilde{p}_i} = 0.$$

The function $\tilde{H} = \tilde{H}(\tilde{q}^1, \ldots, \tilde{q}^n)$ depends only on the $\tilde{q}^i$ and not on $\tilde{p}_i$, and the solutions are given by

$$\tilde{q}^i(t) = a^i, \quad \tilde{p}_i(t) = -\int \frac{\partial \tilde{H}}{\partial \tilde{q}^i}(a^1, \ldots, a^n) dt,$$

where a^i are arbitrary real constants.

Proof Since $\Phi_S^* \omega = \tilde{\omega}$ and $\phi_S^* H = \tilde{H}$ it is clear that Φ_S maps motions of $(M \times U, \tilde{\omega}, \tilde{H})$ to motions of (T^*M, H). In this way we obtain all motions lying in the image of Φ_S. Due to the non-degeneracy of the matrix $\left(\frac{\partial^2 S(x,u)}{\partial x^i \partial u^j} \right)$, the functions $(\tilde{q}^i, \tilde{p}_i)$ form a system of local coordinates, when restricted to an appropriate neighborhood of any given point of $M \times U$. We compute

$$\sum d\tilde{q}^i \wedge d\tilde{p}_i = \sum du^i \wedge d\left(-\frac{\partial S}{\partial u^i} \right) = \sum \frac{\partial^2 S}{\partial x^i \partial u^j} dx^i \wedge du^j.$$

By Lemma 4.9, this is $\tilde{\omega}$. In virtue of the Hamilton–Jacobi equation, for every $u \in U$, the function $x \mapsto \tilde{H}(x, u) = H \circ \Phi_S(x, u) = H \circ dS^u|_x$ is constant, that is, it depends only on u. In the local coordinates $(\tilde{q}^i, \tilde{p}_i)$ this means that $\tilde{H} = \tilde{H}(\tilde{q})$ is a function solely of the $\tilde{q}^i$. So Hamilton's equations reduce to

$$\dot{\tilde{q}}^i = 0, \quad \dot{\tilde{p}}_i = -\frac{\partial \tilde{H}}{\partial \tilde{q}^i}(\tilde{q}),$$

which are trivially solved as indicated. $\square$

Next we discuss an example, cf. [2, Chap. 9, Sect. 47] and references therein. Consider a particle of unit mass moving in the Euclidean plane under the gravitational potential generated by two equal masses placed at distance $2c > 0$, say at the points $f_1 = (c, 0)$ and $f_2 = (-c, 0)$. This problem can be conveniently studied by considering the distances r_1 and r_2 to the points f_1 and f_2, respectively. The gravitational potential is

$$V = -\frac{k}{r_1} - \frac{k}{r_2}$$

for some constant $k > 0$. Away from the Cartesian coordinate axes one can use the globally defined functions

$$\xi := r_1 + r_2, \quad \eta := r_1 - r_2,$$

as coordinates. The level sets of ξ and $|\eta|$ form a *confocal* system of ellipses and hyperbolas, that is a system with common focal points f_1, f_2. From the geometry of conic sections we know that the tangent line at a point q of any ellipse or hyperbola bisects the angle between the two lines connecting q with the foci. This implies that the ellipses and hyperbolas of the confocal system intersect orthogonally. Therefore the Euclidean metric is a linear combination of $d\xi^2$ and $d\eta^2$ (with functions as coefficients).

Proposition 4.11 *In the coordinates (ξ, η) the Euclidean metric takes the form*

$$g = \frac{\xi^2 - \eta^2}{4(\xi^2 - 4c^2)} d\xi^2 + \frac{\xi^2 - \eta^2}{4(4c^2 - \eta^2)} d\eta^2.$$

Proof It is sufficient to check the formula on the ellipses and hyperbolas of the confocal system. The ellipse can be parametrized as $\gamma(t) = (a \cos t, b \sin t)$, where $a^2 - b^2 = c^2$. Then, along the ellipse,

$$r_1^2 = (a \cos t - c)^2 + b^2 \sin^2 t = a^2 \cos^2 t - 2ac \cos t + c^2 + b^2 \sin^2 t$$
$$= c^2 \cos^2 t - 2ac \cos t + a^2 = (a - c \cos t)^2$$

and, thus,

$$r_1 = a - c \cos t, \quad r_2 = a + c \cos t, \quad \xi = 2a, \quad \eta = -2c \cos t.$$

This implies $4c^2 - \eta^2 = 4c^2 \sin^2 t = \dot{\eta}^2$ and, hence,

$$g(\gamma', \gamma') = \frac{\xi^2 - \eta^2}{4(4c^2 - \eta^2)} \dot{\eta}^2 = \frac{1}{4}(\xi^2 - \eta^2) = a^2 - c^2 \cos^2 t = a^2 \sin^2 t + b^2 \cos^2 t.$$

The result coincides with $g_{\text{can}}(\gamma', \gamma')$, the square of the Euclidean length of $\gamma' = (-a \sin t, b \cos t)$. The analogous calculation for the hyperbola is left to the reader (see Appendix A, Exercise 31). This proves that $g_{\text{can}} = g$ on the domain of definition of the coordinates (ξ, η). $\qquad \square$

Corollary 4.12 *In the coordinates (ξ, η), the Hamiltonian of a particle of unit mass moving in the Euclidean plane under the gravitational potential generated by two equal masses placed at $f_1 = (c, 0)$ and $f_2 = (-c, 0)$ has the form*

$$H = 2p_\xi^2 \frac{\xi^2 - 4c^2}{\xi^2 - \eta^2} + 2p_\eta^2 \frac{4c^2 - \eta^2}{\xi^2 - \eta^2} - \frac{4k\xi}{\xi^2 - \eta^2}. \tag{4.2}$$

Proof For any local coordinate system (x^1, x^2) we have

$$H(q, p) = \frac{1}{2} \sum g^{ij}(q) p_i p_j + V(q),$$

where (q^i, p_i) are the corresponding local coordinates on the cotangent bundle of $\mathbb{R}^2$. Inverting the diagonal matrix representing the metric in the coordinates (ξ, η) and dividing by 2 yields the kinetic term in (4.2). The potential term is

$$V = -\frac{k}{r_1} - \frac{k}{r_2} = -\frac{k(r_2 + r_1)}{r_1 r_2} = -\frac{4k\xi}{\xi^2 - \eta^2}.$$

$\square$

In the coordinates (ξ, η), the Hamilton–Jacobi equation $H\left(q, \frac{\partial S}{\partial q}\right) = C = \text{const}$ takes the form

$$2\left(\frac{\partial S}{\partial \xi}\right)^2 (\xi^2 - 4c^2) + 2\left(\frac{\partial S}{\partial \eta}\right)^2 (4c^2 - \eta^2) - 4k\xi = C(\xi^2 - \eta^2).$$

The variables (ξ, p_ξ) and (η, p_η) can be separated as follows. We look for solutions of the system

$$2\left(\frac{\partial S}{\partial \xi}\right)^2 (\xi^2 - 4c^2) - 4k\xi - c_2\xi^2 = c_1,$$

$$2\left(\frac{\partial S}{\partial \eta}\right)^2 (4c^2 - \eta^2) + c_2\eta^2 = -c_1,$$

where c_1, c_2 are constants. Solving each equation for the partial derivatives $\partial S/\partial \xi$ and $\partial S/\partial \eta$, respectively, and integrating yields the following two-parameter family of solutions of the Hamilton–Jacobi equation

$$S(\xi, \eta, c_1, c_2) = \int \sqrt{\frac{c_1 + c_2\xi^2 + 4k\xi}{2(\xi^2 - 4c^2)}} d\xi + \int \sqrt{\frac{-c_1 - c_2\eta^2}{2(4c^2 - \eta^2)}} d\eta.$$

Remark Hamilton–Jacobi theory can be extended to the case of time-dependent Hamiltonians $H(q, p, t)$. The corresponding generalization of the Hamilton–Jacobi equation has the form

$$\frac{\partial S}{\partial t} + H\left(q, \frac{\partial S}{\partial q}, t\right) = 0, \tag{4.3}$$

where $S = S(q, t)$ is now allowed to depend on time t. The choice of the letter S is related to the action functional of a Lagrangian mechanical system as defined in Definition 2.4. Let us very briefly explain this relation. Let $(M, \mathscr{L})$ be a Lagrangian mechanical system with possibly time-dependent Lagrangian. Fix a point $x_0 \in M$

and a time t_0. Suppose that for $(x, t) \in M \times \mathbb{R}$ in a suitable domain there exists a unique motion $s \mapsto \gamma(s)$ (inside a suitable domain in M) such that $\gamma(t_0) = x_0$ and $\gamma(t) = x$. Then we can define

$$S_{(x_0, t_0)}(x, t) := \int_{t_0}^{t} \mathcal{L}(\gamma'(s), s) \, ds.$$

Under appropriate assumptions, it can be shown that $S_{(x_0, t_0)}$ is a smooth family of solutions of the time-dependent Hamilton–Jacobi equation (4.3), see [2, Chap. 9, Sect. 46].

Chapter 5
Classical Field Theory

Abstract A second cornerstone of classical physics besides point-particle mechanics is field theory. Classical field theory is essentially an infinite collection of mechanical systems (one at each point in space) and hence can be viewed as an infinite-dimensional generalization of classical mechanics. More precisely, solutions of classical mechanical systems are smooth curves $t \mapsto \gamma(t)$ from $\mathbb{R}$ to M. In classical field theory, curves from $\mathbb{R}$ are replaced by maps from a higher-dimensional source manifold. In this more general framework we also allow for Lagrangians with explicit time dependence. Another key feature of classical field theory is its manifest incorporation of the laws of Einstein's theory of relativity. This chapter begins with definitions and properties of the central objects, namely fields, Lagrangians, action functionals, and the field-theoretic version of the Euler–Lagrange equations. Modern covariant field theory is customarily formulated in the language of jet bundles, which is also utilized and thus introduced here. We study symmetries and conservation laws of classical field theories in the second part of this chapter, which culminates in the field-theoretic version of Noether's theorem. The penultimate section is devoted to a thorough presentation, from a mathematical perspective, of some prominent examples of classical field theories, such as sigma models, Yang–Mills theory, and Einstein's theory of gravity. A key ingredient of matter-coupled Einstein gravity is the energy-momentum tensor, which is studied in detail in the final section. This chapter is largely based on Ref. [16].

5.1 The Lagrangian, the Action and the Euler–Lagrange Equations

The *fields* of a *classical field theory* are basically smooth maps $f : \mathfrak{S} \to \mathcal{T}$ from a source manifold $\mathfrak{S}$ (possibly with boundary) to a target manifold $\mathcal{T}$. The theory is defined by an action functional of the form

$$S[f] = \int_{\mathfrak{S}} \mathcal{L}(j^k(f)) dvol, \qquad (5.1)$$

© The Author(s) 2017
V. Cortés and A.S. Haupt, *Mathematical Methods of Classical Physics*,
SpringerBriefs in Physics, DOI 10.1007/978-3-319-56463-0_5

where *dvol* is a fixed volume element on $\mathfrak{S}$ and the *Lagrangian* $\mathscr{L}$ is a smooth function on a certain bundle $\text{Jet}^k(\mathfrak{S}, \mathscr{T})$ over $\mathfrak{S}$. The elements of the fiber $\text{Jet}^k_x(\mathfrak{S}, \mathscr{T})$ over $x \in \mathfrak{S}$ are defined as equivalence classes of smooth maps $\mathfrak{S} \to \mathscr{T}$, where two maps are equivalent if their Taylor expansions with respect to some local coordinate system coincide up to order k at x. This condition does not depend on the particular choice of local coordinates. The equivalence class of f with respect to the above relation is denoted by $j^k_x(f) \in \text{Jet}^k_x(\mathfrak{S}, \mathscr{T})$ and is called the *k-th order jet* of f at x. The map

$$j^k(f) : \mathfrak{S} \to \text{Jet}^k(\mathfrak{S}, \mathscr{T}), \quad x \mapsto j^k_x(f),$$

is a smooth section of the *jet bundle* $\text{Jet}^k(\mathfrak{S}, \mathscr{T})$.

In many physically interesting cases the source manifold $\mathfrak{S}$ is interpreted as a space-time of a certain dimension. The Lagrangian mechanical systems discussed so far can be considered as classical field theories of order $k = 1$ with one-dimensional space-time. The source $\mathfrak{S} = I$ is an interval and the target manifold $\mathscr{T}$ is the configuration space of the mechanical system.

For simplicity, we will assume for the moment that $\mathfrak{S}$ is either a bounded domain $\Omega \subset \mathbb{R}^n$ with smooth boundary or $\Omega = \mathbb{R}^n$, and that $\mathscr{T} = \mathbb{R}^m$. (Recall that a domain is a connected open set.) The basic ideas can be explained with almost no loss of generality in this setting, cf. [16]. For the convergence of the action integral one needs to require certain boundary conditions on f, depending on the particular problem under consideration.

Using standard coordinates $(x^1, \ldots, x^n)$ on $\Omega \subset \mathbb{R}^n$ and $(y^1, \ldots, y^m)$ on $\mathbb{R}^m$ we can trivialize the bundle $\text{Jet}^k(\Omega, \mathbb{R}^m)$ and so identify it with $\Omega \times V$, where $V = \text{Jet}^k_0(\mathbb{R}^n, \mathbb{R}^m)$ is the vector space consisting of vector-valued Taylor polynomials of order k. It has natural global coordinates denoted by u^a_I, where $a = 1, \ldots, m$ and $I = (i_1, \ldots, i_\ell) \in \{1, \ldots, n\}^\ell$ runs through all unordered multi-indices of length $\ell \leq k$. So multi-indices which differ only by a permutation are not distinguished. The dimension of V is $m\binom{n+k}{k}$. For any smooth map $f : \Omega \to \mathbb{R}^m$ and multi-index $I = (i_1, \ldots, i_\ell)$ of length $\ell \leq k$ we have

$$u^a_I(j^k(f)) = \partial_I f^a = \partial_{i_1} \cdots \partial_{i_\ell} f^a,$$

where $\partial_i = \partial_{x^i}$ and $f^a = y^a \circ f, a = 1, \ldots, m$. So the k-th jet of f at some point $x \in \Omega$ is simply given x together with the partial derivatives up to order k of the components f^a of f at x.

Example 5.1 Consider the case $n = 3, m = 1$ and $k = 2$. With obvious notational simplifications, the natural coordinates of $V = \text{Jet}^2_0(\mathbb{R}^n, \mathbb{R})$ are

$$u, u_1, u_2, u_3, u_{11}, u_{12}, u_{13}, u_{22}, u_{23}, u_{33}$$

and $\dim V = 10 = \binom{5}{2}$.

In order to derive the equations of motions of a classical field theory it is useful to introduce the *total derivative* $D_i = D_{x^i}$ which is defined by

$$D_i = \partial_i + \sum_I u^a_{I,i} \frac{\partial}{\partial u^a_I},$$

such that, by the chain rule,

$$\partial_i(\mathcal{L}(j^k f)) = (\partial_i \mathcal{L})(j^k f) + \sum_I (\partial_{I,i} f^a) \frac{\partial \mathcal{L}}{\partial u^a_I}(j^k f) = (D_i \mathcal{L})(j^k f).$$

For a multi-index $I = (i_1, \ldots, i_\ell)$ of length $|I| := \ell$ we define

$$D_I = D_{i_1} \cdots D_{i_\ell}$$

and

$$(-D)_I = (-D_{i_1}) \cdots (-D_{i_\ell}) = (-1)^\ell D_I.$$

Theorem 5.2 *Let $\mathcal{L}$ be a smooth function on $\mathrm{Jet}^k(\Omega, \mathbb{R}^m)$ and consider the corresponding action functional (with respect to the canonical volume form). A smooth map $f : \Omega \to \mathbb{R}^m$ is a critical point of the action under smooth variations with compact support in Ω if and only if it has finite action and is a solution of the following system of partial differential equations of order $\leq k$:*

$$\alpha_a := \sum_{|I| \leq k} (-D)_I \frac{\partial \mathcal{L}}{\partial u^a_I}(j^k f) = 0, \quad a = 1, \ldots, m. \tag{5.2}$$

Definition 5.3 The Eqs. (5.2) are called the *Euler–Lagrange equations* or *equations of motion* of the classical field theory defined by the Lagrangian $\mathcal{L}$. A *solution* of the classical field theory is, by definition, a solution of its equations of motion (irrespective of whether it has finite action or not).

In order to compare to the Euler–Lagrange equations in mechanics we observe that

$$\sum_{|I| \leq k} (-D)_I \frac{\partial}{\partial u^a_I} = \frac{\partial}{\partial u^a} + \sum_{1 \leq |I| \leq k} (-D)_I \frac{\partial}{\partial u^a_I}.$$

So for a first order Lagrangian we obtain

$$\frac{\partial}{\partial u^a} - \sum_i D_i \frac{\partial}{\partial u^a_i}.$$

In the case $n = 1$ these are precisely the Euler–Lagrange equations of classical mechanics if we denote $u^a = q^a$, $u^a_1 = \hat{q}^a$, $x^1 = t$, and $D_1 = \frac{d}{dt}$.

Proof (of Theorem 5.2) A smooth map f with finite action is a critical point of the action if and only if for all variations h with compact support we have

$$\frac{d}{d\varepsilon}\bigg|_{\varepsilon=0} S[f + \varepsilon h] = 0.$$

We compute

$$\frac{d}{d\varepsilon}\bigg|_{\varepsilon=0} \mathscr{L}(j^k(f + \varepsilon h)) = \frac{d}{d\varepsilon}\bigg|_{\varepsilon=0} \mathscr{L}(j^k(f) + \varepsilon j^k(h)) = \sum \frac{\partial \mathscr{L}}{\partial u_I^a}(j^k f) u_I^a(j^k(h))$$

$$= \sum \frac{\partial \mathscr{L}}{\partial u_I^a}(j^k f) \partial_I h^a.$$

Applying the divergence theorem to $\partial\Omega$ if $\partial\Omega \neq \emptyset$, or to the boundary of an open ball containing the support of h if $\Omega = \mathbb{R}^n$, we obtain

$$\frac{d}{d\varepsilon}\bigg|_{\varepsilon=0} S[f + \varepsilon h] = \sum \int_\Omega \frac{\partial \mathscr{L}}{\partial u_I^a}(j^k f)\partial_I h^a\, dvol = \sum \int_\Omega (-\partial)_I \left(\frac{\partial \mathscr{L}}{\partial u_I^a}(j^k f)\right) h^a\, dvol$$

$$= \sum \int_\Omega \left((-D)_I \frac{\partial \mathscr{L}}{\partial u_I^a}\right)(j^k f) h^a\, dvol = \sum \int_\Omega \alpha_a h^a\, dvol.$$

This vanishes for all h if and only if $\alpha_a = 0$ for all $a = 1, \ldots, n$.   □

The one-form $\alpha = \sum \alpha_a dy^a$ along f is called the *Euler–Lagrange* one-form. It generalizes the one-form which we encountered in Lagrangian mechanics. We will also consider the *Euler–Lagrange* operators

$$E_a := \sum_{|I| \leq k} (-D)_I \frac{\partial}{\partial u_I^a}, \quad a = 1, \ldots, n, \tag{5.3}$$

which are differential operators acting on smooth functions on the manifold $\mathrm{Jet}^k(\Omega, \mathbb{R}^m)$, such as $\mathscr{L}$. The operator E_a is related to the function $\alpha_a \in C^\infty(\Omega)$ by

$$E_a(\mathscr{L})(j^k f) = \alpha_a.$$

It is clear that the Lagrangian of a classical field theory determines the equations of motion and, hence, the solutions of the theory. Adding a constant to the Lagrangian does not change the equations of motion. More generally, we will show that adding a *total divergence* does not alter the equations of motion.

Definition 5.4 A *total divergence* is a function f on $\mathrm{Jet}^k(\Omega, \mathbb{R}^m)$ of the form

$$f = \mathrm{Div}\, P := \sum_{i=1}^n D_i P^i,$$

where $P = (P^1, \ldots, P^n)$ is a smooth vector-valued function on $\mathrm{Jet}^\ell(\Omega, \mathbb{R}^m)$ for some $\ell \geq k$.

Proposition 5.5 *Let $\mathscr{L} = \mathrm{Div}\, P$ be a total divergence. Then the Euler–Lagrange operators E_a vanish on $\mathscr{L}$:*

$$E_a(\mathscr{L}) = 0.$$

Proof Let $f, h : \Omega \to \mathbb{R}^m$ be smooth maps and assume that h has compact support in Ω. Let Ω' be a domain with smooth boundary, such that Ω' contains the support of h and is relatively compact in Ω. Then, as in the proof of Theorem 5.2, we have that

$$\sum \int_{\Omega} E_a(\mathscr{L}(j^k f)) h^a dvol = \sum \int_{\Omega'} E_a(\mathscr{L}(j^k f)) h^a dvol$$

$$= \sum \int_{\Omega'} \left((-D)_I \frac{\partial \mathscr{L}}{\partial u_I^a} \right) (j^k f) h^a dvol$$

$$= \sum \int_{\Omega'} \frac{\partial \mathscr{L}}{\partial u_I^a} (j^k f) \partial_I h^a dvol$$

$$= \frac{d}{d\varepsilon}\bigg|_{\varepsilon=0} \int_{\Omega'} \mathscr{L}(j^k(f + \varepsilon h)) dvol.$$

By the divergence theorem,

$$\int_{\Omega'} \mathscr{L}(j^k(f + \varepsilon h)) dvol = \int_{\Omega'} \mathrm{Div}\, P(j^{\ell}(f + \varepsilon h)) dvol$$

$$= \int_{\partial \Omega'} \langle P(j^{\ell} f), v \rangle dvol_{\partial \Omega'},$$

where v denotes the outer unit normal and $dvol_{\partial \Omega'}$ the induced volume form of $\partial \Omega' \subset \mathbb{R}^n$. Since the result does not depend on ε, we conclude that $E_a(\mathscr{L}(j^k f))) = 0$ for all smooth maps $f : \Omega \to \mathbb{R}^m$. This proves that $E_a(\mathscr{L}) = 0$. $\qquad\square$

Theorem 5.6 *Let $\mathscr{L}_1, \mathscr{L}_2 \in C^{\infty}(\mathrm{Jet}^k(\mathbb{R}^n, \mathbb{R}^m))$ be two Lagrangians of order $\leq k$ defined on $\mathbb{R}^n$. Then $\mathscr{L}_1$ and $\mathscr{L}_2$ have the same Euler–Lagrange equations if and only if $\mathscr{L}_1 - \mathscr{L}_2$ is a total divergence.*

Proof By the previous proposition, we already know that $\mathscr{L}_1$ and $\mathscr{L}_2$ have the same Euler–Lagrange equations if $\mathscr{L}_1 - \mathscr{L}_2$ is a total divergence. Therefore it suffices to show that a Lagrangian $\mathscr{L} \in C^{\infty}(\mathrm{Jet}^k(\mathbb{R}^n, \mathbb{R}^m))$ such that $E_a(\mathscr{L}) = 0$ for all a (that is a Lagrangian with trivial equations of motion) is necessarily a total divergence. For every $f \in C^{\infty}(\mathbb{R}^n, \mathbb{R}^m)$ we have

$$\mathscr{L}(j^k f) = \int_0^1 \frac{d}{d\varepsilon} \mathscr{L}(\varepsilon j^k f) d\varepsilon - \mathscr{L}(s_0), \tag{5.4}$$

where s_0 stands for the zero section of the vector bundle $\mathrm{Jet}^k(\mathbb{R}^n, \mathbb{R}^m) \to \mathbb{R}^n$, that is the k-th order jet of the constant map $\mathbb{R}^n \to \mathbb{R}^m$, $x \mapsto 0$. Clearly the function $\mathscr{L}(s_0) \in C^{\infty}(\mathbb{R}^n)$ does not depend on f and can be written in the form

$\mathcal{L}(s_0(x)) = \partial_1 F_1(x)$ for some function $F_1 \in C^\infty(\mathbb{R}^n)$. This shows that it is a total divergence $\sum D_i F_i$, since we can simply put $F_i = 0$ for $2 \le i \le n$. It remains to show that the integral on the right-hand side of (5.4) is also a total divergence. By the chain rule we have

$$\frac{d}{d\varepsilon}\mathcal{L}(\varepsilon j^k f) = \sum \partial_I f^a \frac{\partial \mathcal{L}}{\partial u_I^a}(\varepsilon j^k f).$$

Performing partial integrations for each term $\ell = |I| \le k$ brings this to the form

$$\sum f^a (-D)_I \frac{\partial \mathcal{L}}{\partial u_I^a}(\varepsilon j^k f) + (\operatorname{Div} P_\varepsilon)(j^{2k} f), \tag{5.5}$$

where $P_\varepsilon = (P_\varepsilon^1, \ldots, P_\varepsilon^n)$ is a vector-valued function on $\operatorname{Jet}^{2k-1}(\mathbb{R}^n, \mathbb{R}^m)$ depending smoothly on all the variables, including the parameter ε. (Notice that $\operatorname{Div} P_\varepsilon$ is thus a function on $\operatorname{Jet}^{2k}(\mathbb{R}^n, \mathbb{R}^m)$ depending smoothly on all the variables, including the parameter ε.) The first term in (5.5) vanishes by the assumption $E_a(\mathcal{L}) = \sum (-D)_I \frac{\partial \mathcal{L}}{\partial u_I^a} = 0$. Thus,

$$\int_0^1 \frac{d}{d\varepsilon}\mathcal{L}(\varepsilon j^k f) d\varepsilon = \left(\operatorname{Div} \int_0^1 P_\varepsilon d\varepsilon\right)(j^{2k} f)$$

is a total divergence.   □

5.2 Automorphisms and Conservation Laws

We begin by observing that for every pair of smooth manifolds M, N the group $\operatorname{Diff}(M) \times \operatorname{Diff}(N)$ acts naturally on $C^\infty(M, N)$. In fact, given a group element $g = (\varphi, \psi)$ and a smooth map $f : M \to N$ we have

$$g \cdot f = \psi \circ f \circ \varphi^{-1}.$$

This action induces an action on the jet bundle $\operatorname{Jet}^k(M, N)$:

$$g \cdot (j_x^k f) := j_{\varphi(x)}^k (g \cdot f), \quad x \in M.$$

Definition 5.7 Let M, N be smooth manifolds, $dvol$ a volume form on M, $n = \dim M$ and $\mathcal{L} \in C^\infty(\operatorname{Jet}^k(M, N))$ a Lagrangian. An element $g = (\varphi, \psi) \in \operatorname{Diff}^+(M) \times \operatorname{Diff}(N)$ is called an *automorphism* of the Lagrangian n-form $\mathcal{L} dvol$ if for all $f \in C^\infty(M, N), x \in M$:

$$\mathcal{L}(g \cdot j_x^k f)(\varphi^* dvol)_x = \mathcal{L}(j_x^k f) dvol_x. \tag{5.6}$$

Here $\text{Diff}^+(M) \subset \text{Diff}(M)$ denotes the subgroup of orientation preserving diffeomorphisms of M.

Pulling back the n-forms in Eq. (5.6) by φ^{-1} we obtain the equivalent equation

$$\mathscr{L}(g \cdot j^k_{\varphi^{-1}(x)} f) dvol_x = \mathscr{L}(j^k_{\varphi^{-1}(x)} f)((\varphi^{-1})^* dvol)_x. \tag{5.7}$$

Proposition 5.8 *If $g \in \text{Diff}^+(M) \times \text{Diff}(N)$ is an automorphism of $\mathscr{L} dvol$, then*

$$S[g \cdot f] = \int_M \mathscr{L}(j^k(g \cdot f)) dvol = S[f] = \int_M \mathscr{L}(j^k f) dvol$$

for all $f \in C^\infty(M, N)$.

Proof This follows from

$$\left[(\varphi^{-1})^* \left(\mathscr{L}(j^k f) dvol\right)\right]_x = \mathscr{L}(j^k_{\varphi^{-1}(x)} f)((\varphi^{-1})^* dvol)_x \overset{(5.6)}{=} \mathscr{L}(g \cdot j^k_{\varphi^{-1}(x)} f) dvol_x$$
$$= \mathscr{L}(j^k_x(g \cdot f)) dvol_x.$$

$\square$

The calculation in this proof shows that $g = (\varphi, \psi) \in \text{Diff}^+(M) \times \text{Diff}(N)$ is an automorphism of $\mathscr{L} dvol$ if and only if for all $f \in C^\infty(M, N)$:

$$(\varphi^{-1})^* \left(\mathscr{L}(j^k f) dvol\right) = \mathscr{L}(j^k(g \cdot f)) dvol,$$

or, equivalently,

$$\varphi^* \left(\mathscr{L}(j^k(g \cdot f)) dvol\right) = \mathscr{L}(j^k f) dvol.$$

Definition 5.9 A *conservation law* for a Lagrangian $\mathscr{L} \in C^\infty(\text{Jet}^k(\mathbb{R}^n, \mathbb{R}^m))$ is a total divergence Div P which vanishes on all solutions of the Euler–Lagrange equations of $\mathscr{L}$.

Theorem 5.10 (Noether) *Consider a classical field theory defined by a Lagrangian $\mathscr{L} \in C^\infty(\text{Jet}^k(\mathbb{R}^n, \mathbb{R}^m))$ and denote by dvol the standard volume form of $\mathbb{R}^n$. With every local one-parameter group of local automorphisms of $\mathscr{L} dvol$ one can associate a conservation law of the form Div $P = \sum Q^a E_a(\mathscr{L})$, where $Q^a = Y^a - \sum u_i^a X^i$ is determined by the vector field $\sum_{i=1}^n X^i \frac{\partial}{\partial x^i} + \sum_{a=1}^m Y^a \frac{\partial}{\partial y^a}$ generating the local one-parameter group.*

For the proof of Noether's theorem we will use a series of lemmas. Let X, Y be smooth vector fields on $\mathbb{R}^n$ and $\mathbb{R}^m$, respectively. Then the flow of $Z = X + Y$ on $\mathbb{R}^n \times \mathbb{R}^m$ induces a flow on $\text{Jet}^k(\mathbb{R}^n, \mathbb{R}^m)$ and we denote the corresponding vector field on $\text{Jet}^k(\mathbb{R}^n, \mathbb{R}^m)$ by $\text{pr}^{(k)} Z$.

Definition 5.11 The vector field $\text{pr}^{(k)} Z$ is called the k-th *prolongation* of Z.

Lemma 5.12 *Consider a Lagrangian* $\mathcal{L} \in C^\infty(\mathrm{Jet}^k(\mathbb{R}^n, \mathbb{R}^m))$ *and the standard volume form dvol on* $\mathbb{R}^n$. *Let* X, Y *be smooth vector fields on* $\mathbb{R}^n$ *and* $\mathbb{R}^m$, *respectively. Then* $Z = X + Y$ *is an infinitesimal automorphism of* $\mathcal{L}$ *dvol if and only if*

$$(\mathrm{pr}^{(k)} Z)(\mathcal{L}) + \mathcal{L} \mathrm{div} X = 0. \tag{5.8}$$

Proof Recall that the divergence $\mathrm{div} X$ of a smooth vector field X on $\mathbb{R}^n$ is characterized by the equation

$$L_X dvol = \mathrm{div} X dvol,$$

where L_X denotes the Lie derivative. The differential equation characterizing infinitesimal automorphisms is obtained by differentiating the Eq. (5.6) with respect to t after evaluation on a one-parameter group $g_t = (\varphi_t, \psi_t) \in \mathrm{Diff}^+(\mathbb{R}^n) \times \mathrm{Diff}(\mathbb{R}^m)$. The derivative of the left-hand side is

$$\left. \frac{d}{dt} \right|_{t=0} \mathcal{L}(g_t \cdot j_x^k f)(\varphi_t^* dvol)_x = (\mathrm{pr}^{(k)} Z)(\mathcal{L})|_{j_x^k f} dvol_x + \mathcal{L}(j_x^k f)(\mathrm{div} X)_x dvol_x,$$

whereas the right-hand side does not depend on t.   □

Lemma 5.13 *Let* X, Y *be smooth vector fields on* $\mathbb{R}^n$ *and* $\mathbb{R}^m$, *respectively. Then the k-th prolongation of* $Z = X + Y$ *is given by*

$$\mathrm{pr}^{(k)} Z = Z + \sum_{1 \leq |J| \leq k} \sum_a Y_J^a \frac{\partial}{\partial u_J^a}, \tag{5.9}$$

where

$$Y_J^a = D_J Q^a + \sum u_{J,i}^a X^i, \quad Q^a = Y^a - \sum u_i^a X^i.$$

Proof Since X and Y commute, the flow φ_t^Z of $Z = X + Y$ decomposes as the composition $\varphi_t^Z = \varphi_t^X \circ \varphi_t^Y$ of the flows of its summands. This implies that $\tilde{\varphi}_t^Z = \tilde{\varphi}_t^X \circ \tilde{\varphi}_t^Y$, where $\tilde{\varphi}_t^X, \tilde{\varphi}_t^Y, \tilde{\varphi}_t^Z$ denote the induced flows on $\mathrm{Jet}^k(\mathbb{R}^n, \mathbb{R}^m)$. Differentiation with respect to t yields that

$$\mathrm{pr}^{(k)} Z = \mathrm{pr}^{(k)} X + \mathrm{pr}^{(k)} Y.$$

Therefore, it suffices to check the formula (5.9) in the special cases $X = 0$ and $Y = 0$.

We first consider the case $X = 0$. Then $Q^a = Y^a$ and $Y_J^a = D_J Y^a$. So, what we have to show is

$$\mathrm{pr}^{(k)} Y = Y + \sum_{1 \leq |J| \leq k} \sum_a D_J Y^a \frac{\partial}{\partial u_J^a}. \tag{5.10}$$

Let us denote by ψ_t the flow of Y on $\mathbb{R}^m$ and by $g_t = (\mathrm{Id}, \psi_t)$ the corresponding local one-parameter group of local diffeomorphisms of $\mathbb{R}^n \times \mathbb{R}^m$. The induced action on $\mathrm{Jet}^k(\mathbb{R}^n, \mathbb{R}^m)$ is given by

$$g_t \cdot j_x^k f = j_x^k(\psi_t \circ f) = \left(x, \psi_t(f(x)), \{\partial_J(\psi_t^a \circ f)(x)\}_{1 \leq |J| \leq k, \, a=1,\ldots,m}\right)$$

for all $f \in C^\infty(\mathbb{R}^n, \mathbb{R}^m)$ and $x \in \mathbb{R}^n$, where $\psi_t^a = y^a \circ \psi_t$. Differentiating this equation with respect to t yields

$$\mathrm{pr}^{(k)}Y|_{j_x^k f} = \left(0, Y_{f(x)}, \{\partial_J(Y^a \circ f)(x)\}_{1 \leq |J| \leq k, \, a=1,\ldots,m}\right)$$
$$= \left(0, Y_{f(x)}, \{(D_J Y^a)(j_x^k f)\}_{1 \leq |J| \leq k, \, a=1,\ldots,m}\right).$$

This is precisely the right-hand side of (5.10).

Next we assume $Y = 0$, in which case $Q^a = -\sum u_i^a X^i$. Firstly, we consider the case $k = 1$ as a warm up. Now

$$Y_j^a = D_j Q^a + \sum u_{i,j}^a X^i = -\sum u_i^a \partial_j X^i.$$

So, what we have to show is

$$\mathrm{pr}^{(1)}X = X - \sum u_i^a \partial_j X^i \frac{\partial}{\partial u_j^a}. \tag{5.11}$$

Let us denote by φ_t the flow of X and by $g_t = (\varphi_t, \mathrm{Id})$ the corresponding local one-parameter group of local diffeomorphisms of $\mathbb{R}^n \times \mathbb{R}^m$. It acts on $\mathrm{Jet}^1(\mathbb{R}^n, \mathbb{R}^m)$ by

$$g_t \cdot j_x^k f = j_{\varphi_t(x)}^k(f \circ \varphi_t^{-1}) = (\varphi_t(x), f(x), \{\partial_j(f^a \circ \varphi_t^{-1})|_{\varphi_t(x)}\}_{j,a}).$$

Differentiation with respect to t yields

$$\mathrm{pr}^{(1)}X|_{j_x^k f} = \frac{d}{dt}\bigg|_{t=0} g_t \cdot j_x^k f = \left(X_x, 0, \left\{\frac{d}{dt}\bigg|_{t=0} \partial_j(f^a \circ \varphi_t^{-1})|_{\varphi_t(x)}\right\}\right).$$

We compute

$$d(f \circ \varphi_t^{-1})|_{\varphi_t(x)} = df \circ d(\varphi_t^{-1})|_{\varphi_t(x)} = \left[(\varphi_t^{-1})^* df\right]_{\varphi_t(x)}.$$

In order to differentiate this with respect to t, we consider the time-dependent matrix-valued function $F_t(x) := F(t, x) := (\varphi_t^{-1})^* df|_x$. So, what we have to differentiate is $F(t, \varphi_t(x))$. Its total time-derivative is

$$\frac{d}{dt}\bigg|_{t=0} F(t, \varphi_t(x)) = \frac{\partial F}{\partial t}(0, x) + dF_0 X_x = -L_X df|_x + (\mathrm{Hess}\, f)(X_x, \cdot)$$
$$= -d(df X)|_x + (\mathrm{Hess}\, f)(X_x, \cdot) = -df \circ dX|_x.$$

This shows that

$$\frac{d}{dt}\Big|_{t=0} (\partial_j (f^a \circ \varphi_t^{-1})|_{\varphi_t(x)}) = -df^a \circ dX|_x \partial_j = -\sum \partial_i f^a \partial_j X^i(x)$$

$$= -\left(\sum u_i^a \partial_j X^i\right)(j_x^1 f),$$

proving (5.11) in the case $k = 1$.

Let us finally consider the case of general $k \geq 1$. We first claim that for every multi-index of length $\leq k$ we have

$$(D_J Q^a)(j^{k+1} f) = -\partial_J X(f^a). \tag{5.12}$$

In fact,

$$Q^a(j^k f) = -\sum (u_i^a X^i)(j^k f) = -\sum \partial_i f^a X^i = -X(f^a)$$

implies (5.12). As a consequence,

$$Y_J^a(j^{k+1} f) = \left(D_J Q^a + \sum u_{J,i}^a X^i\right)(j^{k+1} f) = -\partial_J X(f^a) + \sum (\partial_{J,i} f^a) X^i.$$

So, what we have to show is

$$\frac{d}{dt}\Big|_{t=0} (\partial_J (f^a \circ \varphi_t^{-1})|_{\varphi_t(x)}) = -\partial_J X(f^a)|_x + \sum (\partial_{J,i} f^a) X^i|_x. \tag{5.13}$$

To calculate the left-hand side we put $F_t(x) := F(t, x) := \partial_J (f^a \circ \varphi_t^{-1})_x$. Then we see that

$$\frac{d}{dt}\Big|_{t=0} (\partial_J (f^a \circ \varphi_t^{-1})|_{\varphi_t(x)}) = \frac{d}{dt}\Big|_{t=0} F(t, \varphi_t(x)) = \frac{\partial F}{\partial t}(0, x) + dF_0 X_x$$

$$= -\partial_J X(f^a)|_x + d(\partial_J f^a) X|_x$$

This coincides with the right-hand side of (5.13).   □

Now we can prove Noether's theorem.

Proof (of Theorem 5.10) We have to show that $\sum Q^a E_a(\mathcal{L})$ is a total divergence. By partial integration we have

$$\sum Q^a E_a(\mathcal{L}) = \sum Q^a (-D)_J \frac{\partial \mathcal{L}}{\partial u_J^a} = \sum (D_J Q^a) \frac{\partial \mathcal{L}}{\partial u_J^a} + \text{Div } V$$

for some vector valued function $V = (V^1, \ldots, V^n)$ on $\text{Jet}^{2k-1}(\mathbb{R}^n, \mathbb{R}^m)$. It suffices to show that $\sum (D_J Q^a) \frac{\partial \mathcal{L}}{\partial u_J^a}$ is a total divergence. Recall that we defined $Y_J^a = D_J Q^a + \sum u_{J,i}^a X^i$ for $1 \leq |J| \leq k$. The formula holds also for $|J| = 0$:

$$Y^a = Q^a + \sum u_i^a X^i.$$

Using this as well as Lemmas 5.13 and 5.12 we have

$$\sum (D_J Q^a) \frac{\partial \mathscr{L}}{\partial u_J^a} = \sum (Y_J^a - \sum u_{J,i}^a X^i) \frac{\partial \mathscr{L}}{\partial u_J^a} = (-X + \mathrm{pr}^{(k)} Z) \mathscr{L} - \sum u_{J,i}^a X^i \frac{\partial \mathscr{L}}{\partial u_J^a}$$

$$= -\sum X^i D_i \mathscr{L} + (\mathrm{pr}^{(k)} Z) \mathscr{L} = -\sum X^i D_i \mathscr{L} - \mathscr{L} \operatorname{div} X$$

$$= -\sum D_i (\mathscr{L} X^i),$$

which is a total divergence.   □

In the proof of Theorem 5.10 we have shown that $\sum Q^a E_a(\mathscr{L})$ is the total divergence of the vector-valued function with components $P^i = V^i - \mathscr{L} X^i$, where $X = \sum X^i \partial_i$ is the projection of the infinitesimal automorphism Z onto the source manifold $\mathbb{R}^n$ and V^i is a function on $\mathrm{Jet}^{2k-1}(\mathbb{R}^n, \mathbb{R}^m)$ obtained by partial integration. This means that $P = (P^1, \ldots, P^n)$ can be explicitly computed by performing the partial integrations. In the case $k = 1$ we obtain the following result.

Corollary 5.14 *Let* $\mathscr{L} \in C^\infty(\mathrm{Jet}^1(\mathbb{R}^n, \mathbb{R}^m))$ *be a first order Lagrangian. Then Noether's conservation law associated with an infinitesimal automorphism* $Z = X + Y$ *as in Theorem 5.10 is the total divergence of the vector-valued function* $P = (P^1, \ldots, P^n)$ *on* $\mathrm{Jet}^1(\mathbb{R}^n, \mathbb{R}^m)$ *given by*

$$P^i = -\sum Q^a \frac{\partial \mathscr{L}}{\partial u_i^a} - \mathscr{L} X^i, \quad Q^a = Y^a - \sum u_j^a X^j.$$

Proof It suffices to observe that $V^i := -\sum Q^a \frac{\partial \mathscr{L}}{\partial u_i^a}$ satisfies

$$-\sum Q^a D_i \frac{\partial \mathscr{L}}{\partial u_i^a} = \sum (D_i Q^a) \frac{\partial \mathscr{L}}{\partial u_i^a} + \sum D_i V^i.$$

□

As another corollary we obtain the generalization to time-dependent Lagrangian mechanical systems on $\mathbb{R}^m$ of Noether's theorem (Theorem 2.12), which concerned time-independent mechanical systems.

Corollary 5.15 *Let* $\mathscr{L} = \mathscr{L}(t, q^1, \ldots, q^m, \hat{q}^1, \ldots, \hat{q}^m)$ *be a time-dependent Lagrangian mechanical system on* $\mathbb{R}^m$. *Then every infinitesimal automorphism* $Z = X + Y$, $X = X^1 \partial_t \in \mathfrak{X}(\mathbb{R})$, $Y = \sum Y^a \partial_{q^a} \in \mathfrak{X}(\mathbb{R}^m)$, *of* $\mathscr{L} dt$ *gives rise to an integral of motion*

$$f = -P^1 = \sum Q^a \frac{\partial \mathscr{L}}{\partial \hat{q}^a} + \mathscr{L} X^1, \quad Q^a = Y^a - \sum \hat{q}^a X^1.$$

Example 5.16 Notice that in the case $X = 0$, an infinitesimal automorphism $Y \in \mathfrak{X}(\mathbb{R}^m)$ of $\mathscr{L}dt$ is the same as an infinitesimal automorphism of $\mathscr{L}$ and the integral of motion takes the familiar form of Theorem 2.12:

$$f = \sum Y^a \frac{\partial \mathscr{L}}{\partial \hat{q}^a} = d\mathscr{L} Y^{\text{ver}}.$$

The prolongation of a vector field $Y \in \mathfrak{X}(\mathbb{R}^m)$ given in (5.10) simplifies in the considered case $k = 1 = n$ as

$$\text{pr}^{(1)} Y = Y + \sum D_t Y^a \frac{\partial}{\partial \hat{q}^a}$$

and thus Y is an infinitesimal automorphism if and only if

$$Y(\mathscr{L}) + \sum D_t Y^a \frac{\partial \mathscr{L}}{\partial \hat{q}^a} = 0.$$

Evaluating this along the velocity vector field γ' of a smooth curve $\gamma : I \to \mathbb{R}^m$, we obtain

$$\sum Y^a(t) \frac{\partial \mathscr{L}(\gamma'(t))}{\partial q^a} + \sum \dot{Y}^a(t) \frac{\partial \mathscr{L}(\gamma'(t))}{\partial \hat{q}^a} = 0,$$

where $Y^a(t) := Y^a(\gamma(t))$. So Y is an infinitesimal automorphism if and only if the latter equation holds for all γ.

Next we state Noether's theorem for time-dependent Lagrangian mechanical systems in its general form replacing $\mathbb{R}^m$ by an arbitrary smooth manifold.

Theorem 5.17 *Let $\mathscr{L} \in C^\infty(\mathbb{R} \times TM)$ be a time-dependent Lagrangian mechanical system on a smooth manifold M. Then every infinitesimal automorphism $Z = X + Y$, $X = X^1 \partial_t \in \mathfrak{X}(\mathbb{R})$, $Y \in \mathfrak{X}(M)$, of $\mathscr{L}dt$ gives rise to an integral of motion*

$$f = Y^{\text{ver}}(\mathscr{L}) + (\mathscr{L} - \xi(\mathscr{L}))X^1, \tag{5.14}$$

where $\xi \in \mathfrak{X}(TM)$ denotes as usual the vertical vector field generated by scalar multiplication $(s, v) \mapsto e^s v$ in the fibers of TM and Y^{ver} denotes the vertical lift of Y.

Proof Let (U, φ) be a local chart of M. According to Corollary 5.15, we have the following integral of motion on TU:

$$f_{(U,\varphi)} = \sum Q^a \frac{\partial \mathscr{L}}{\partial \hat{q}^a} + \mathscr{L}X^1, \quad Q^a = Y^a - \sum \hat{q}^a X^1,$$

where Y^a, $a = 1, \ldots, m$, $m = \dim M$, are the components of Y with respect to the local chart $(q^1, \ldots, q^m) = \varphi$ and $(t, q^1, \ldots, q^m, \hat{q}^1, \ldots, \hat{q}^m)$ are the corresponding local coordinates of $\mathbb{R} \times TM = \mathrm{Jet}^1(\mathbb{R}, M)$, defined on $\mathbb{R} \times TU = \mathrm{Jet}^1(\mathbb{R}, U)$. Since $\sum Y^a \frac{\partial}{\partial \hat{q}^a} = Y^{\mathrm{ver}}|_U$ and $\sum \hat{q}^a \frac{\partial}{\partial \hat{q}^a} = \xi|_U$, we see that $f_{(U, \varphi)}$ is the restriction of the globally defined (and manifestly[1] coordinate independent) function f. This shows that f is an integral of motion. □

Example 5.18 Consider the special case when $Y = 0$. The expression for the prolongation of $X = X^1 \partial_t \in \mathbb{R}$ on $\mathrm{Jet}^1(\mathbb{R}, M)$ in local coordinates $(t, q, \hat{q})$ follows immediately from (5.11):

$$\mathrm{pr}^{(1)} X = X - \sum \hat{q}^a \partial_t X^1 \frac{\partial}{\partial \hat{q}^a}. \tag{5.15}$$

The vector field $\mathrm{pr}^{(1)} X$ is in fact coordinate independent and obviously coincides with $X - \partial_t X^1 \xi$. We conclude that X is an infinitesimal automorphism of $\mathscr{L} dt$ if and only if

$$0 = X(\mathscr{L}) - \partial_t X^1 \xi(\mathscr{L}) + \mathscr{L} \partial_t X^1 = X(\mathscr{L}) + (\mathscr{L} - \xi(\mathscr{L})) \partial_t X^1.$$

The right-hand side vanishes, in particular, if $\mathscr{L}$ is time-independent and $X = \partial_t$. The corresponding integral of motion (5.14) is precisely

$$\mathscr{L} - \xi(\mathscr{L}) = -E,$$

the energy, up to the factor -1. So we see that the energy is the integral of motion associated with the invariance of the Lagrangian under translations in time.

The notion of an infinitesimal automorphism of a Lagrangian n-form can be generalized as follows.

Definition 5.19 Let X and Y be smooth vector fields on $\mathbb{R}^n$ and $\mathbb{R}^m$, respectively, $\mathscr{L} \in C^\infty(\mathrm{Jet}^k(\mathbb{R}^n, \mathbb{R}^m))$ a Lagrangian and $d\mathrm{vol}$ the standard volume form of $\mathbb{R}^n$. We say that $Z = X + Y$ is an *infinitesimal automorphism of* $\mathscr{L} d\mathrm{vol}$ *up to a divergence* if

$$(\mathrm{pr}^{(k)} Z)(\mathscr{L}) + \mathscr{L} \operatorname{div} X$$

is a total divergence.

Noether's theorem can be easily generalized as follows (see Appendix A, Exercise 40).

Theorem 5.20 (Noether) *Under the above assumptions, let* $Z = X + Y$ *be an infinitesimal automorphism of* $\mathscr{L} d\mathrm{vol}$ *up to a divergence. Then*

$$\sum Q^a E_a(\mathscr{L}), \quad \text{defined by} \quad Q^a = Y^a - \sum u_i^a X^i,$$

---

[1] Note that in the definition of f, we did not use any coordinates.

is a total divergence, where X^i and Y^a are the components of $X \in \mathfrak{X}(\mathbb{R}^n)$ and $Y \in \mathfrak{X}(\mathbb{R}^m)$.

5.3 Why are Conservation Laws called Conservation Laws?

Consider a classical field theory defined by a Lagrangian n-form $\mathscr{L} dvol$, where $\mathscr{L} \in C^\infty(\mathrm{Jet}^k(\mathbb{R}^n, \mathbb{R}^m))$ and $dvol$ is the standard volume form on $\mathbb{R}^n$. We denote the standard coordinates of $\mathbb{R}^n$ by $(x^0, \ldots x^{n-1})$ and think of $t = x^0$ as the time-coordinate and of $\mathbf{x} = (x^1, \ldots, x^{n-1})$ as the spatial position. Suppose that we are given a conservation law $\mathrm{Div}\, P$, $P \in C^\infty(\mathrm{Jet}^\ell(\mathbb{R}^n, \mathbb{R}^m), \mathbb{R}^n)$, $\ell \geq k$. We denote by

$$J(t, \mathbf{x}) := P(j_x^\ell f),$$

the evaluation of P on a solution $f \in C^\infty(\mathbb{R}^n, \mathbb{R}^m)$.

Definition 5.21 The vector-valued function

$$J = (J^0, \mathbf{J}) : \mathbb{R}^n \to \mathbb{R}^n = \mathbb{R} \times \mathbb{R}^{n-1}$$

is called the *current*, the function J^0 is called the *charge density*, $\mathbf{J}$ is called the *flux density* and the spatial integral

$$Q(t) := \int_{\mathbb{R}^{n-1}} J^0(t, \mathbf{x}) dx^1 \wedge \cdots \wedge dx^{n-1} \tag{5.16}$$

is called the *charge*.

    If P is associated with an infinitesimal automorphism Z of $\mathscr{L} dvol$, as described in Theorem 5.10 (or, more generally, in Theorem 5.20), then J is called the *Noether current* and Q is called the *Noether charge* associated with $-Z$.

The sign is included here such that the notation of Theorem 5.10 is consistent with the usual conventions for the Noether charge in the physics literature. With this notation, the Noether current of an infinitesimal automorphism $Z = X + Y$ of a first order Lagrangian $\mathscr{L} dvol$ on $\mathbb{R}^n$ is given by

$$J^i = \sum_a \left(Y^a \circ f - \sum_j \partial_j f^a X^j \right) \frac{\partial \mathscr{L}}{\partial u_i^a}(j^1 f) + \mathscr{L}(j^1 f) X^i, \tag{5.17}$$

see Corollary 5.14. Note that in the case $n = 1$, the current reduces to the charge density, $J = J^0 = Q$, and as we already observed, the equation $\mathrm{Div}\, P = 0$ reduces to the statement that P is an integral of motion, that is to $Q'(t) = 0$ (for all solutions f). This is generalized in the next theorem.

Theorem 5.22 *Suppose that the current falls off sufficiently fast at infinity, in the sense that for every bounded interval I there exists a Lebesgue integrable function $F : \mathbb{R}^{n-1} \to [0, \infty]$ such that*

$$|J^0(t, \mathbf{x})| \le F(\mathbf{x}),$$
$$\left|\frac{\partial J^0}{\partial t}(t, \mathbf{x})\right| \le F(\mathbf{x}),$$

for all $(t, \mathbf{x}) \in I \times \mathbb{R}^{n-1}$ and

$$\lim_{r \to \infty} \int_{\partial B_r(0)} \langle \mathbf{J}, v \rangle dvol_{\partial B_r(0)} = 0,$$

where v denotes the outer unit normal of the sphere $\partial B_r(0) \subset \mathbb{R}^{n-1}$. Then the charge is conserved, that is $t \mapsto Q(t)$ is constant.

Proof In virtue of the first inequality, we see that the total charge (5.16) is finite by Lebesgue's theorem. The second inequality then shows that Q is differentiable and

$$Q'(t) = \int_{\mathbb{R}^{n-1}} \frac{\partial J^0}{\partial t}(t, \mathbf{x}) d^{n-1}x,$$

where we have abbreviated $d^{n-1}x = dx^1 \wedge \cdots \wedge dx^{n-1}$. Since

$$\operatorname{div} J = \partial_t J^0 + \operatorname{div}_{\mathbb{R}^{n-1}} \mathbf{J} = 0,$$

we can rewrite this as

$$Q'(t) = -\int_{\mathbb{R}^{n-1}} \operatorname{div}_{\mathbb{R}^{n-1}} \mathbf{J}(t, \mathbf{x}) d^{n-1}x = -\lim_{r \to \infty} \int_{B_r(0)} \operatorname{div}_{\mathbb{R}^{n-1}} \mathbf{J}(t, \mathbf{x}) d^{n-1}x$$
$$= -\lim_{r \to \infty} \int_{\partial B_r(0)} \langle \mathbf{J}, v \rangle dvol_{\partial B_r(0)},$$

and the resulting limit is zero by the last assumption of the theorem. $\qquad \square$

5.4 Examples of Field Theories

Here we discuss some examples of classical field theories. The examples play important roles in active research on contemporary theories of high-energy particle physics, such as the standard model of particle physics and string theory.

5.4.1 Sigma Models

Let (M, g) and (N, h) be pseudo-Riemannian manifolds of dimension m and n, respectively. We assume that (N, h) is oriented and denote its volume form by $dvol_h$. The most natural[2] first order Lagrangian n-form $\mathcal{L}dvol_h$ for maps $f \in C^\infty(N, M)$ is given by

$$\mathcal{L}(j^1 f) = \frac{1}{2} \langle df, df \rangle, \tag{5.18}$$

where $\langle \cdot, \cdot \rangle = \langle \cdot, \cdot \rangle_{h,g}$ is the (possibly indefinite) fiber-wise scalar product on $T^*N \otimes f^*TM$ induced by h and g. In local coordinates (x^i) in a neighborhood of $x \in N$ and (y^a) in a neighborhood of $f(x) \in M$ we have

$$df_x = \sum \frac{\partial f^a(x)}{\partial x^i} dx^i|_x \otimes \left. \frac{\partial}{\partial y^a} \right|_{f(x)}$$

and

$$\langle df_x, df_x \rangle = \sum g_{ab}(f(x)) h^{ij}(x) \frac{\partial f^a(x)}{\partial x^i} \frac{\partial f^b(x)}{\partial x^j},$$

where (h^{ij}) denotes the inverse matrix of (h_{ij}).

Theorem 5.23 *The Euler–Lagrange equations for the Lagrangian n-form $\mathcal{L}dvol_h$ given by (5.18) are equivalent to*

$$\tau(f) := \operatorname{tr}_h \nabla df = 0, \tag{5.19}$$

where ∇ is the connection in the vector bundle $T^*N \otimes f^*TM$ induced by the Levi-Civita connections in TN and TM.

Proof See Appendix A, Exercise 39. $\square$

Definition 5.24 A smooth map $f : N \to M$ which is a solution of Eq. (5.19) is called a *harmonic map*. The vector field $\tau(f)$ along f is called the *tension* of f.

Example 5.25 (*Harmonic functions*) Consider the case when the target manifold (M, g) is simply the Euclidean line $\mathbb{R}$. Then f^*TM can be identified with the trivial line bundle over N and

$$\tau(f) = \Delta f := \operatorname{div} \operatorname{grad} f$$

is given by the *Laplace operator* Δ associated with the pseudo-Riemannian metric h on the source manifold N. The solutions of the equation $\Delta f = 0$ are called *harmonic functions*. Recall that

$$\operatorname{grad} f = h^{-1} df$$

---

[2]This is to be understood in the sense that it generalizes the kinetic term in the standard Lagrangian (2.4) of classical mechanics.

and

$$\operatorname{div} X = \operatorname{tr} \nabla X$$

for every vector field X on (N, h). So indeed

$$\operatorname{div} \operatorname{grad} f = \operatorname{tr} \nabla (h^{-1} df) = \operatorname{tr}_h \nabla df = \tau(f).$$

Here we have used that $\nabla_v (h^{-1} df) = h^{-1} \nabla_v df$ for every $v \in TN$. Note that the pseudo-Riemannian Laplace equation $\Delta f = 0$ is linear, whereas the harmonic map equation (5.19) is in general non-linear, since the connection ∇ depends on f. Observe that in the special case when (N, h) is a pseudo-Euclidean vector space we can write the metric as $h = \sum \varepsilon_i (dx^i)^2$, where $\varepsilon_i \in \{\pm 1\}$, and $\Delta f = \sum \varepsilon_i \partial_i^2$.

Example 5.26 Generalizing the previous example, we consider the case when the target manifold (M, g) is a pseudo-Euclidean vector space. Then the components $\tau^a(f)$ of the tension $\tau(f)$ with respect to an affine coordinate system on the target manifold M are given by

$$\tau^a(f) = \Delta f^a.$$

So a smooth map f from a pseudo-Riemannian manifold to a pseudo-Euclidean vector space is harmonic if and only if its components f^a are harmonic functions.

In the physics literature the Lagrangian (5.18) is called a *sigma-model* and in the case of Example 5.26 it is called a *linear sigma model*, since the equations of motion are linear. The components of the map $f : N \to M$ are considered as scalar fields. We can enlarge the class of sigma-model Lagrangians by including a potential:

$$\mathcal{L}(j^1 f) = \frac{1}{2} \langle df, df \rangle - V(f),$$

where $V \in C^\infty(M)$.

Example 5.27 (*Geodesics as harmonic maps*) Consider the case when the source manifold (N, h) is the Euclidean line $\mathbb{R}$. Then a harmonic map $f : \mathbb{R} \to (M, g)$ is the same as a geodesic.

Proposition 5.28 *The Lie group* $\operatorname{Isom}^+(N, h) \times \operatorname{Isom}(M, g)$ *consists of automorphisms of the sigma-model Lagrangian n-form* $\mathcal{L} dvol_h$ *defined by* (5.18).

Proof This follows from the fact that the group $\operatorname{Isom}^+(N, h)$ of orientation preserving isometries of (N, h) preserves the metric volume form $dvol_h$ and that $\operatorname{Isom}(N, h) \times \operatorname{Isom}(M, g)$ preserves the Lagrangian $\mathcal{L}$. $\qquad\square$

5.4.2 Pure Yang–Mills Theory

Let E be a real or complex vector bundle of rank k over an oriented pseudo-Riemannian manifold (M, g) with a *reduction of the structure group* of E to some compact[3] subgroup $G \subset \mathrm{GL}(k, \mathbb{K})$, where $\mathbb{K} = \mathbb{R}$ or $\mathbb{C}$. Such a G-*reduction* in the vector bundle E is defined as a principal G-subbundle $\mathscr{F}_G(E) \subset \mathscr{F}(E)$ of the bundle of frames of E. The elements of $\mathscr{F}_G(E)$ are called G-*frames*.

Example 5.29 An $\mathrm{O}(k)$-reduction in a real vector bundle E is equivalent to a (positive definite fiber) metric h in E. The corresponding principal $\mathrm{O}(k)$-subbundle of the frame bundle $\mathscr{F}(E)$ is the bundle of orthonormal frames in the metric vector bundle (E, h). A $\mathrm{U}(k)$-reduction in a complex vector bundle E is equivalent to a Hermitian metric h in E. The corresponding principal $\mathrm{U}(k)$-subbundle of the frame bundle $\mathscr{F}(E)$ is the bundle of unitary frames in the Hermitian vector bundle (E, h).

Note that since G is compact every G-reduction induces a metric h in E if $\mathbb{K} = \mathbb{R}$ and a Hermitian metric in E if $\mathbb{K} = \mathbb{C}$. A connection ∇ in a vector bundle E endowed with a G-reduction $\mathscr{F}_G(E)$ is called a G-*connection* if the parallel transport with respect to ∇ maps G-frames to G-frames. We denote the curvature of ∇ by

$$F = F^{\nabla} \in \Gamma(\Lambda^2 T^*M \otimes \mathfrak{g}(E)),$$

where $\mathfrak{g}(E_x)$ denotes the Lie algebra of the subgroup $G(E_x) \subset \mathrm{GL}(E_x)$ of elements preserving the set of G-frames in $E_x, x \in M$. For convenience, we will write $\mathfrak{so}(E)$, $\mathfrak{su}(E)$, $\mathfrak{gl}(E)$ etc. rather than $\mathfrak{so}(k, \mathbb{R})(E)$, $\mathfrak{su}(k)(E)$, $\mathfrak{gl}(k, \mathbb{K})(E)$ etc.

Example 5.30 An $\mathrm{O}(k)$-connection in a metric vector bundle (E, h) of rank k is the same as a metric connection. A $\mathrm{U}(k)$-connection in a Hermitian vector bundle (E, h) of rank k is the same as a Hermitian connection.

Given a vector bundle E with structure group G over a pseudo-Riemannian manifold (M, g), the space of fields of *pure Yang–Mills theory* is the affine space $\mathscr{A}_G(E)$ of G-connections in E. By choosing a reference connection ∇^0 we can write $\nabla = \nabla^0 + \Phi$, where $\Phi : M \to T^*M \otimes \mathfrak{g}(E)$ is a smooth section. In this way we can identify the affine space $\mathscr{A}_G(E)$ with the vector space $\Gamma(T^*M \otimes \mathfrak{g}(E))$. Using this identification, one can define jets of ∇ by considering jets of the map $\Phi = \nabla - \nabla^0$. The Yang–Mills Lagrangian is

$$\mathscr{L}(j^1(\nabla)) = -\frac{1}{2}\langle F, F \rangle, \tag{5.20}$$

where $\langle \cdot, \cdot \rangle$ is the (fiber-wise) scalar product on $\Lambda^2 T^*M \otimes \mathfrak{g}(E)$ obtained as the tensor product of the scalar product $\langle \cdot, \cdot \rangle_\Lambda$ on $\Lambda^2 T^*M$ induced by g and the scalar product $\langle A, B \rangle_\mathfrak{g} = -\mathrm{tr}\,(AB)$ on $\mathfrak{g}(E)$. Notice that $\langle A, A \rangle_\mathfrak{g} = -\mathrm{tr}\,A^2 = \mathrm{tr}\,AA^{\dagger} \geq 0$, where

---

[3]Recall that a compact subgroup of a Lie group is automatically a Lie subgroup.

$A^{\dagger}$ denotes the adjoint of A with respect to the G-invariant metric h. In particular, $\langle\cdot,\cdot\rangle_{\mathfrak{g}}$ is real valued, irrespective of whether E is a real or a complex vector bundle. The scalar product $\langle\cdot,\cdot\rangle_{\Lambda}$ is normalized such that $\alpha\wedge\beta$ is of unit length if α, β are orthonormal.

Recall that every connection satisfies the *Bianchi identity*

$$d^{\nabla}F = 0, \tag{5.21}$$

where d^{∇} is the covariant exterior derivative acting on differential forms with values in $\mathrm{End}(E)\supset\mathfrak{g}(E)$.

Theorem 5.31 *The Euler–Lagrange equations for the Yang–Mills Lagrangian* (5.20) *are equivalent to*

$$d^{\nabla}*F = 0. \tag{5.22}$$

Before we begin the proof of this proposition let us first discuss an example.

Example 5.32 (Maxwell theory) Let us consider the special case when $(M, g) = (\mathbb{R}^4, dt^2 - \sum_{\alpha=1}^{3}(dx^{\alpha})^2)$ is the four-dimensional Minkowski space and E is a Hermitian line bundle. So the structure group is the Abelian group $G = U(1)$. Identifying the Lie algebra $\mathfrak{u}(1) = \sqrt{-1}\mathbb{R} \cong \mathbb{R}$ with $\mathbb{R}$, we can consider F as a real-valued 2-form. The Bianchi and Yang–Mills equations reduce to $dF = d*F = 0$. So the differential form F is closed and co-closed, which implies the second order equations $*d*dF = 0$ and $d*d*F = 0$. The spatial components $F_{\alpha\beta} = F(\partial_{\alpha}, \partial_{\beta})$, $\alpha,\beta\in\{1,2,3\}$, define a real-valued time-dependent 2-form in $\mathbb{R}^3$. Identifying $\mathbb{R}^3 \cong \Lambda^2(\mathbb{R}^3)^*$ by means of contraction of a vector with the Euclidean volume form, this 2-form defines a time-dependent vector field $\mathbf{B}$ in $\mathbb{R}^3$. The remaining components $E^{\alpha} := F_{0\alpha} = F(\partial_t, \partial_{\alpha})$, $\alpha = 1, 2, 3$, define a time-dependent vector field $\mathbf{E} = \sum_{\alpha} E^{\alpha}\partial_{\alpha}$ in $\mathbb{R}^3$. The vector fields $\mathbf{E}$ and $\mathbf{B}$ can be interpreted as the electric and the magnetic field in Maxwell's theory of electromagnetism and the Yang–Mills equation reduces to half of Maxwell's equations in the vacuum, that is

$$\mathrm{div}\,\mathbf{E} = 0, \quad \mathrm{rot}\,\mathbf{B} = \frac{\partial}{\partial t}\mathbf{E},$$

whereas the Bianchi identity reduces to the other half of Maxwell's vacuum equations, that is

$$\mathrm{div}\,\mathbf{B} = 0, \quad \mathrm{rot}\,\mathbf{E} = -\frac{\partial}{\partial t}\mathbf{B},$$

see Appendix A, Exercise 43.

In order to derive the Yang–Mills equation (5.22) it is helpful to first rewrite the Yang–Mills Lagrangian as:

$$-\frac{1}{2}\langle F, F\rangle dvol_g = -\frac{1}{2}\langle F\wedge *F\rangle_{\mathfrak{g}}, \tag{5.23}$$

where $\langle \cdot \wedge \cdot \rangle_{\mathfrak{g}}$ denotes the $\Lambda^n T^* M$-valued pairing of $\Lambda^\ell T^* M \otimes \mathfrak{g}(E)$ with $\Lambda^{n-\ell} T^* M \otimes \mathfrak{g}(E)$ obtained by combining the wedge product and the scalar product on $\mathfrak{g}(E)$. Here $n = \dim M$ and $\ell = 2$, but in later calculations we will also need the case $\ell = 1$. This is a consequence of the following lemma.

Lemma 5.33 *For every section α of $\Lambda^2 T^* M \otimes \mathfrak{g}(E)$ we have*

$$\langle \alpha, \alpha \rangle dvol_g = \langle \alpha \wedge *\alpha \rangle_{\mathfrak{g}}. \tag{5.24}$$

Proof Let $(e_i)_{i=1,\dots,n}$ be a local frame of TM and put $\alpha_{ij} := \alpha(e_i, e_j)$. Then we can write

$$\alpha = \frac{1}{2} \sum \alpha_{ij} e^{ij},$$

where (e^i) is the dual local frame of $T^* M$ and we have abbreviated $e^i \wedge e^j =: e^{ij}$. Now we can compute

$$
\begin{aligned}
\langle \alpha \wedge *\alpha \rangle_{\mathfrak{g}} &= \frac{1}{4} \sum \langle \alpha_{ij}, \alpha_{k\ell} \rangle_{\mathfrak{g}} e^{ij} \wedge *e^{k\ell} \\
&= \frac{1}{4} \sum \langle \alpha_{ij}, \alpha_{k\ell} \rangle_{\mathfrak{g}} \langle e^{ij}, e^{k\ell} \rangle_{\Lambda} dvol_g \\
&= \langle \alpha, \alpha \rangle dvol_g.
\end{aligned}
$$

$\square$

Proposition 5.34 *Let (e_i) be any local frame in TM and (e^i) the dual frame in $T^* M$. Then the Yang–Mills Lagrangian is locally given by*

$$-\frac{1}{2} \langle F, F \rangle = \frac{1}{4} \sum \mathrm{tr} \, (F_{ij} F^{ij}),$$

where $F_{ij} = F(e_i, e_j)$, $F^{ij} = \sum g^{ii'} g^{jj'} F_{i'j'}$, $g^{ij} = g^{-1}(e^i, e^j)$.

Proof Choosing an orthonormal frame (e_i), the calculation in the proof of the previous lemma shows that

$$
\begin{aligned}
-\frac{1}{2} \langle F, F \rangle &= -\frac{1}{8} \sum \langle F_{ij}, F_{k\ell} \rangle_{\mathfrak{g}} \langle e^{ij}, e^{k\ell} \rangle_{\Lambda} \\
&= -\frac{1}{4} \sum \langle F_{ij}, F_{ij} \rangle_{\mathfrak{g}} \langle e^{ij}, e^{ij} \rangle_{\Lambda} = -\frac{1}{4} \sum \langle F_{ij}, F^{ij} \rangle_{\mathfrak{g}} = \frac{1}{4} \sum \mathrm{tr} \, (F_{ij} F^{ij}).
\end{aligned}
$$

It is easy to check that the expression $\sum F_{ij} F^{ij}$ is independent of the choice of frame. This proves that the formula $-\frac{1}{2} \langle F, F \rangle = \frac{1}{4} \sum \mathrm{tr} \, (F_{ij} F^{ij})$ holds in every local frame.
$\square$

As a corollary we obtain the usual expression for the Yang–Mills Lagrangian in the physics literature.

Corollary 5.35 *In local coordinates (x^μ), $\mu = 1, \ldots, n$, the Yang–Mills Lagrangian on (M, g) is given by*

$$-\frac{1}{2}\langle F, F \rangle = \frac{1}{4}\sum \operatorname{tr}(F_{\mu\nu}F^{\mu\nu}),$$

where $F_{\mu\nu} = F(\partial_\mu, \partial_\nu)$, $F^{\mu\nu} = \sum g^{\mu\mu'}g^{\nu\nu'}F_{\mu'\nu'}$, $g^{\mu\nu} = g^{-1}(dx^\mu, dx^\nu)$.

Lemma 5.36 *Let ∇ be a G-connection in E and*

$$\alpha \in \Omega^1(\mathfrak{g}(E)) = \Gamma(T^*M \otimes \mathfrak{g}(E))$$

a one-form with values in $\mathfrak{g}(E)$. Then

$$F^{\nabla+\alpha} = F^\nabla + d^\nabla\alpha + \alpha \wedge \alpha,$$

where $(\alpha \wedge \alpha)(X, Y) := \alpha(X)\alpha(Y) - \alpha(Y)\alpha(X) = [\alpha(X), \alpha(Y)]$ for all $X, Y \in \mathfrak{X}(M)$.

Proof See Appendix A, Exercise 44. □

Corollary 5.37 *Let ∇ be a G-connection in E, $\alpha \in \Omega^1(\mathfrak{g}(E))$, and consider the curvature F_ε of the one-parameter family of G-connections $\nabla^\varepsilon = \nabla + \varepsilon\alpha$. Then*

$$\left.\frac{d}{d\varepsilon}\right|_{\varepsilon=0} \langle F_\varepsilon \wedge *F_\varepsilon \rangle_\mathfrak{g} = 2\langle d^\nabla\alpha \wedge *F \rangle_\mathfrak{g},$$

where $F = F_0$.

Proof Since $F_\varepsilon = F + \varepsilon d^\nabla\alpha + \varepsilon^2\alpha \wedge \alpha$, we have

$$\frac{1}{2}\left.\frac{d}{d\varepsilon}\right|_{\varepsilon=0} \langle F_\varepsilon \wedge *F_\varepsilon \rangle_\mathfrak{g} = \frac{1}{2}\left.\frac{d}{d\varepsilon}\right|_{\varepsilon=0} \langle F_\varepsilon, F_\varepsilon \rangle dvol_g = \langle d^\nabla\alpha, F \rangle dvol_g.$$

□

Proof (of Theorem 5.31) Using the preceding notation we compute

$$-\left.\frac{d}{d\varepsilon}\right|_{\varepsilon=0} \mathcal{L}(j^1(\nabla^\varepsilon))dvol_g = \langle d^\nabla\alpha \wedge *F \rangle_\mathfrak{g} = -\operatorname{tr}(d^\nabla\alpha \wedge *F)$$

$$= -\operatorname{tr}\left(d^\nabla(\alpha \wedge *F) + \alpha \wedge d^\nabla * F\right)$$

$$= -\operatorname{tr} d^\nabla(\alpha \wedge *F) + \langle \alpha \wedge d^\nabla * F \rangle_\mathfrak{g}.$$

Since

$$\operatorname{tr} d^\nabla(\alpha \wedge *F) = d \operatorname{tr}(\alpha \wedge *F),$$

we conclude that for variations α with compact support we have

$$-\int_M \frac{d}{d\varepsilon}\Big|_{\varepsilon=0} \mathscr{L}(j^1(\nabla^\varepsilon))dvol_g = \int_M \langle \alpha \wedge d^\nabla * F \rangle_{\mathfrak{g}} dvol_g.$$

This proves that the equations of motion of pure Yang–Mills theory are indeed $d^\nabla * F = 0$. $\square$

Gauge Transformations

Given a vector bundle $E \to M$, we denote by $\mathrm{Aut}(E)$ its group of automorphisms. If the vector bundle is equipped with a G-reduction $\mathscr{F}_G(E) \subset \mathscr{F}(E)$, then the subgroup of $\mathrm{Aut}(E)$ consisting of those automorphisms which preserve the reduction will be denoted by $\mathrm{Aut}_G(E)$. It coincides with the group of sections of the group bundle $G(E) \subset \mathrm{GL}(E)$.

Definition 5.38 $\mathrm{Aut}_G(E)$ is called the *gauge group* of the vector bundle E with G-reduction $\mathscr{F}_G(E)$. The elements of $\mathrm{Aut}_G(E)$ are called *gauge transformations*.

Gauge transformations play a similar role in the study of vector bundles to diffeomorphisms in the study of manifolds. The group $\mathrm{Aut}(E)$ acts $C^\infty(M)$-linearly on the space of smooth sections $\Gamma(E)$ by

$$\mathrm{Aut}(E) \times \Gamma(E) \to \Gamma(E), \quad (\varphi, s) \to \varphi s.$$

Proposition 5.39 *The action of* $\mathrm{Aut}_G(E)$ *on* $\Gamma(E)$ *induces an affine action on the space* $\mathscr{A}_G(E)$ *of* G-connections.

Proof The transformation of $\mathscr{A}_G(E)$ induced by $\varphi \in \mathrm{Aut}_G(E)$ is

$$\nabla \mapsto \nabla' = \nabla^\varphi := \varphi \circ \nabla \circ \varphi^{-1}. \tag{5.25}$$

To see that it is affine, we apply ∇' to a section $s \in \Gamma(E)$:

$$\nabla' s = \varphi(\nabla \varphi^{-1} s) = \varphi \nabla(\varphi^{-1}) s + \nabla s.$$

So

$$\nabla' = \nabla + \varphi \nabla(\varphi^{-1}) \tag{5.26}$$

is related to ∇ by a translation. $\square$

Proposition 5.40 *Let* ∇ *be a connection in a vector bundle* $E \to M$ *and* $\varphi \in \mathrm{Aut}(E)$ *an automorphism. Then the curvature* F' *of the gauge transformed connection* $\nabla' = \nabla^\varphi$ *is related to the curvature* F *of* ∇ *by the natural* $C^\infty(M)$-*linear action of* $\mathrm{Aut}(E)$ *on* $\Omega^2(\mathrm{End}(E))$:

$$F' = \varphi \circ F \circ \varphi^{-1}.$$

Proof This is a straightforward consequence of (5.25) and the definition of the curvature. $\square$

Corollary 5.41 *The Yang–Mills Lagrangian* (5.20) *is invariant under* $\mathrm{Aut}(E)$.

Proof This follows from the fact that the trace form $(A, B) \mapsto \mathrm{tr}(AB)$ on $\mathfrak{gl}(E)$, which was used in the definition of $\langle F, F \rangle$, is invariant under conjugation. $\qquad \square$

This clarifies the role of gauge transformations in Yang–Mills theory. For that reason, Yang–Mills theory is also known as *gauge theory*. In particular in the physics literature, often a further distinction is made between Abelian and non-Abelian gauge theories depending on the property of the respective gauge group. In this context, only the latter is customarily referred to as Yang–Mills theory. This additional distinction is used not only for historical reasons but also because the corresponding physical theories differ significantly in both cases [9, 12, 15].

Remark Let us denote by $\mathscr{A}_G^{YM}(E) \subset \mathscr{A}_G(E)$ the subset consisting of *Yang–Mills connections*, that is solutions to the Yang–Mills equations. It follows from Corollary 5.41 that the gauge group $\mathrm{Aut}_G(E)$ acts on $\mathscr{A}_G^{YM}(E)$. The quotient

$$\mathscr{A}_G^{YM}(E)/\mathrm{Aut}_G(E) \tag{5.27}$$

is the set of equivalence classes of Yang–Mills connections. Notice that the space of connections as well as the group of gauge transformations are of infinite dimension. Nevertheless it is possible to prove that under suitable extra assumptions (on the signature of the metric g, the connections, and gauge transformations), quotients similar to (5.27) are sometimes finite-dimensional smooth manifolds encoding differential topological information about the smooth manifold M. These problems are studied in mathematical work on gauge theory. A particularly successful instance is *Donaldson theory* (see, for example, Ref. [4]), which is concerned with four-dimensional smooth manifolds M, and restricts attention to the class of self-dual or anti-self-dual connections (with respect to a Riemannian metric g), that is connections satisfying the *self-duality equation*

$$*F = F$$

or the *anti-self-duality equation*

$$*F = -F.$$

In virtue of the Bianchi identity, each of these equations implies the Yang–Mills equation:

$$d^\nabla * F = \pm d^\nabla F = 0.$$

The Connection One-Form

Let ∇ be a connection in a vector bundle $E \to M$ of rank k. Given a local frame $f = (f_1, \ldots, f_k)$ of E, defined on some open subset $U \subset M$, there exists a system of one-forms A_j^i, $i, j \in \{1, \ldots, k\}$, such that

$$\nabla_X f_i = \sum A_i^j(X) f_j \tag{5.28}$$

for all vector fields X on $U \subset M$. The matrix-valued one-form

$$A = A^{\nabla, f} = (A^i_j)_{i, j \in \{1, \dots, k\}}$$

is called the *connection one-form* of ∇ with respect to the local frame f. The Eq. (5.28) can be written in matrix-notation as

$$\nabla f = fA.$$

Solving it with respect to A we obtain

$$A = f^{-1} \nabla f,$$

where $f^{-1} = (f_1^*, \dots, f_k^*)^\mathsf{T}$ is the column vector which is the transposed of the dual frame $f^* = (f_1^*, \dots, f_n^*)$. The notation is consistent with the interpretation of a frame $b = (b_1, \dots, b_k)$ of E at $x \in M$ (such as $b = f_x$) as an isomorphism

$$b : \mathbb{K}^k \to E_x, \quad v \to bv = \sum v^i b_i,$$

where bv is the product of the E_x-valued row vector $b = (b_1, \dots, b_k)$ with the column vector $v = (v^1, \dots, v^k)^\mathsf{T}$. The inverse of this map is given by

$$E_x \to \mathbb{K}^k, \quad w \mapsto (b_1^*(w), \dots, b_k^*(w))^\mathsf{T}.$$

Proposition 5.42 *Let ∇ be a connection and f a local frame in a vector bundle E over $\mathbb{K} \in \{\mathbb{R}, \mathbb{C}\}$. The connection form $A' := A^{\nabla', f}$ of a gauge transformed connection $\nabla' = \nabla^\varphi$, $\varphi \in \mathrm{Aut}(E)$, is related to the connection form $A = A^{\nabla, f}$ of ∇ by*

$$A' = \psi A \psi^{-1} + \psi d(\psi^{-1}), \tag{5.29}$$

where $\psi = f^{-1} \circ \varphi \circ f : U \to \mathrm{GL}(k, \mathbb{K})$ is the matrix-valued function representing the automorphism φ in the frame f.

Proof Using that $\varphi = f \circ \psi \circ f^{-1}$ we compute

$$\alpha := \nabla' - \nabla \overset{(5.26)}{=} \varphi \nabla(\varphi^{-1}) = f \psi f^{-1} \left[(\nabla f) \psi^{-1} f^{-1} + f \nabla(\psi^{-1} f^{-1}) \right]$$
$$= f \psi A \psi^{-1} f^{-1} + f \psi \nabla(\psi^{-1} f^{-1}) = f \psi A \psi^{-1} f^{-1} + f \psi d(\psi^{-1}) f^{-1} + f \nabla(f^{-1})$$

and, hence,

$$A' - A = f^{-1} \alpha(f) = \psi A \psi^{-1} + \psi d(\psi^{-1}) + \underbrace{\nabla(f^{-1}) f}_{= -f^{-1} \nabla f = -A}.$$

$\square$

Proposition 5.43 *With the notation of the previous proposition we have*

$$A^{\nabla',f} = A^{\nabla,f'},$$

where $f' = \varphi^{-1} \circ f = f \circ \psi^{-1}$.

Proof From the formula $A^{\nabla,f} = f^{-1}\nabla f$ we can immediately deduce the general transformation rule

$$A^{\varphi\circ\nabla\circ\varphi^{-1},\varphi\circ f} = A^{\nabla,f}. \tag{5.30}$$

Substituting f' for f in this formula, we obtain the claim. $\square$

The last proposition shows that a gauge transformation $\nabla \mapsto \nabla' = \nabla^\varphi = \varphi \circ \nabla \circ \varphi^{-1}$ by φ has the same effect on the connection form as a change of frame $f \mapsto f' = \varphi^{-1} \circ f = f \circ \psi^{-1}$ by the inverse transformation φ^{-1}, or equivalently, by the corresponding change of basis ψ^{-1} in $\mathbb{K}^k$.

Substituting a one-parameter group $\psi_s : U \to GL(k, \mathbb{K})$ into (5.29) and differentiating with respect to s, we obtain the local expression for the action on a connection ∇ of an infinitesimal gauge transformation represented by $\tau : U \to \mathfrak{gl}(k, \mathbb{K})$:

$$A \mapsto [\tau, A] - d\tau.$$

To apply Noether's theorem (Theorem 5.10) and to compute the Noether current (5.17) corresponding to an infinitesimal gauge transformation, one can use the local trivialization f of the bundle E and a system of coordinates (x^μ), $\mu = 1, \ldots, n$, on $U \subset M$ to describe a connection ∇ by its connection form $A = \sum A_\mu dx^\mu = \sum A^i_{j\mu} dx^\mu \otimes e^*_j \otimes e_i$, which we consider as a map

$$U \to U \times ((\mathbb{R}^n)^* \otimes \mathfrak{gl}(k, \mathbb{K})), \quad x \mapsto (x, (A^i_{j\mu}(x))).$$

Here $(e_1, \ldots, e_k)$ denotes the standard basis of $\mathbb{K}^k$. Now one can compute the corresponding Noether current by the formula

$$J^\nu = \sum_a Y^a(A)\frac{\partial\mathscr{L}}{\partial u^a_\nu}(j^1 A),$$

where a is an index numerating the elements of the basis $(dx^\mu \otimes e^*_j \otimes e_i)$ of $V := (\mathbb{R}^n)^* \otimes \mathfrak{gl}(k, \mathbb{K})$ such that

$$(u^a(j^1 A)) = (u^i_{j\mu}(j^1 A)) = (A^i_{j\mu}) \quad \text{and} \quad (u^a_\nu(j^1 A)) = (u^i_{j\mu,\nu}(j^1 A)) = (\partial_\nu A^i_{j\mu}).$$

The components $(Y^a) = (Y^i_{j\mu})$ of the infinitesimal automorphism

$$\mathfrak{X}(U \times V) \ni Y = \sum Y^a \frac{\partial}{\partial u^a} : U \times V \to V,$$

$$(x, A) \mapsto Y(x, A) = [\tau(x), A] - d\tau(x)$$

are given by

$$Y^i_{j\mu}(x, A) = \sum \tau^i_{j'}(x) A^{j'}_{j\mu} - A^i_{j'\mu} \tau^{j'}_j(x) - \partial_\mu \tau^i_j(x), \quad A = (A^i_{j\mu}).$$

In order to obtain an explicit expression for J^ν it suffices to calculate the partial derivatives

$$\frac{\partial \mathscr{L}}{\partial u^a_\nu}(j^1 A) = \frac{\partial \mathscr{L}}{\partial u^i_{j\mu,\nu}}(j^1 A) = (F^{\mu\nu})^j_i,$$

where $((F^{\mu\nu})^i_j)_{i,j\in\{1,\dots,k\}}$ is the matrix representing the endomorphism $F^{\mu\nu}$ with respect to the local frame f. So we obtain:

Proposition 5.44 *The Noether current associated with an infinitesimal gauge transformation represented by $\tau : U \to \mathfrak{gl}(k, \mathbb{K})$ is given by*

$$J^\mu = \operatorname{tr} \sum F^{\mu\nu}([\tau, A_\nu] - \partial_\nu \tau).$$

5.4.3 The Einstein–Hilbert Lagrangian

In theories of gravity the space-time metric is no longer fixed but is a *dynamical* field of the theory, that is subject to equations of motion. The most important Lagrangian for a pseudo-Riemannian metric g on a smooth manifold M is the *Einstein–Hilbert Lagrangian* n-form

$$\mathscr{L}(j^2 g) = \frac{1}{16\pi\kappa} scal \, dvol_g \,, \tag{5.31}$$

where $scal$ is the scalar curvature of (M, g) and κ is the gravitational coupling constant. It describes pure gravity, that is, gravity in the absence of matter fields. More complicated theories of gravity can be obtained by including matter fields described, for instance, by adding to the Einstein–Hilbert term a sigma-model Lagrangian (for a map from M to some target manifold) or a Yang–Mills Lagrangian (for a connection in some vector bundle $E \to M$). Notice that the fields of the resulting theory are coupled through the common dependence on the space-time metric, in the sense that the resulting equations of motion generally form a coupled system of partial differential equations.

Theorem 5.45 *The Euler–Lagrange equations for the Einstein–Hilbert Lagrangian are equivalent to the* Einstein vacuum equation:

$$Ric - \frac{1}{2} scal \cdot g = 0, \tag{5.32}$$

which is equivalent to Ric = 0 if n = dim M ≥ 3 and is satisfied for every pseudo-Riemannian metric g if n ≤ 2.

Proof Let g_ε be a smooth family of pseudo-Riemannian metrics on M such that the symmetric tensor field

$$h := \frac{\partial}{\partial \varepsilon}\bigg|_{\varepsilon=0} g_\varepsilon \qquad (5.33)$$

has compact support. In order to compute the derivative of $\mathcal{L}(j^2 g_\varepsilon)$ with respect to ε, we first observe that

$$\frac{\partial}{\partial \varepsilon}\bigg|_{\varepsilon=0} dvol_{g_\varepsilon} = \frac{1}{2} \operatorname{tr}(g^{-1}h) dvol_g, \qquad (5.34)$$

where $g = g_0$. The derivative of the curvature tensor R_ε of the Levi-Civita connection ∇^ε of the metric g_ε is expressed in terms of the tensor field

$$\alpha := \frac{\partial}{\partial \varepsilon}\bigg|_{\varepsilon=0} \nabla^\varepsilon = \frac{\partial}{\partial \varepsilon}\bigg|_{\varepsilon=0} (\nabla^\varepsilon - \nabla) \in \Omega^1(\operatorname{End} TM)$$

by

$$\frac{\partial}{\partial \varepsilon}\bigg|_{\varepsilon=0} R_\varepsilon = d^\nabla \alpha, \quad \nabla = \nabla^0,$$

as follows easily from Lemma 5.36. As a consequence, the Ricci curvature Ric_ε of g_ε can be computed as follows:

$$\frac{\partial}{\partial \varepsilon}\bigg|_{\varepsilon=0} Ric_\varepsilon(X, Y) = \frac{\partial}{\partial \varepsilon}\bigg|_{\varepsilon=0} \operatorname{tr}(Z \mapsto R_\varepsilon(Z, X)Y) = \operatorname{tr}(Z \mapsto d^\nabla\alpha(Z, X)Y),$$

where $X, Y, Z \in \mathfrak{X}(M)$. Omitting Z we can write this in terms of a local frame (e_i) of TM as

$$\frac{\partial}{\partial \varepsilon}\bigg|_{\varepsilon=0} Ric_\varepsilon(X, Y) = \sum e^i \left[(\nabla_{e_i}\alpha)_X Y - (\nabla_X \alpha)_{e_i} Y \right], \qquad (5.35)$$

where $\alpha_X = \alpha(X)$ denotes the endomorphism field obtained by evaluating the one-form α on the vector field X. We have used that $d^\nabla\alpha(X, Y)Z = (\nabla_X\alpha)_Y Z - (\nabla_Y\alpha)_X Z$ for all $X, Y, Z \in \mathfrak{X}(M)$. For the scalar curvature $scal_\varepsilon$ of g_ε we obtain

$$\frac{\partial}{\partial \varepsilon}\bigg|_{\varepsilon=0} scal_\varepsilon = \frac{\partial}{\partial \varepsilon}\bigg|_{\varepsilon=0} (\operatorname{tr} g_\varepsilon^{-1} Ric_\varepsilon) = -\operatorname{tr}(g^{-1}hg^{-1}Ric)$$

$$+ \operatorname{tr}\underbrace{\left(g^{-1} \frac{\partial}{\partial \varepsilon}\bigg|_{\varepsilon=0} Ric_\varepsilon \right)}_{=:f}, \qquad (5.36)$$

where Ric denotes the Ricci curvature of g. If we can show that f is a total divergence, then (5.34) and (5.36) imply that

$$\int_M \frac{\partial}{\partial \varepsilon}\Big|_{\varepsilon=0} scal_\varepsilon \, dvol_{g_\varepsilon} = -\int_M \mathrm{tr}\left(g^{-1}h\left(g^{-1}Ric - \frac{scal}{2}\mathrm{Id}\right)\right) dvol_g.$$

From this formula we see that the equations of motion are equivalent to (5.32). Taking the trace of $g^{-1}Ric - \frac{scal}{2}\mathrm{Id}$, we see that every solution has $scal = 0$ if $n \neq 2$, which implies $Ric = 0$ by (5.32). To compute f at some point $p \in M$ we can assume that $\nabla_{e_i} e_j = 0$ at p. Then from (5.35) we compute at p:

$$f = \sum g^{jk} e^i \left[(\nabla_{e_i}\alpha)_{e_j} e_k - (\nabla_{e_j}\alpha)_{e_i} e_k \right]$$
$$= \sum e^i \nabla_{e_i}(g^{jk}\alpha_{e_j} e_k) - \sum g^{jk} e_k \nabla_{e_j}(e^i \circ \alpha_{e_i}),$$

where $g^{jk} = g^{-1}(e^j, e^k)$. The first sum is the divergence of the vector field $\mathrm{tr}_g\, \alpha$ obtained by contraction of the first two factors of $T^*M \otimes T^*M \otimes TM$ with the help of the metric. The other sum is the divergence of the vector field $v = g^{-1}\sum e^i \circ \alpha_{e_i}$ obtained by contraction of the first and last factors by duality and identification of the remaining factor T^*M with TM using the metric:

$$\sum g^{jk} e_k \nabla_{e_j}(e^i \circ \alpha_{e_i}) = \sum g^{jk} e_k \nabla_{e_j} gv = \sum g^{jk} g(\nabla_{e_j} v, e_k) = \sum e^j \nabla_{e_j} v = \mathrm{div}\, v.$$

So we have proven that f is the divergence of the vector field $\mathrm{tr}_g\, \alpha - v$, which does not depend on the particular frame. $\qquad\square$

Let (M, g, or) be an oriented pseudo-Riemannian manifold and denote by $dvol_{g,or}$ its metric volume form with respect to the orientation or. In local coordinates $(x^1, \ldots, x^n)$ on some open set $U \subset M$, we have

$$dvol_{g,or}|_U = \varepsilon \sqrt{|\det(g_{ij})|} dx^1 \wedge \cdots \wedge dx^n, \quad \varepsilon \in \{1, -1\}, \tag{5.37}$$

if $dx^1 \wedge \cdots \wedge dx^n \in \varepsilon \cdot or$, where $g_{ij} = g(\partial_i, \partial_j)$.

Proposition 5.46 *For all* $\varphi \in \mathrm{Diff}(M)$, *we have*

$$scal_{\varphi^*g} dvol_{\varphi^*g, \varphi^* or} = \varphi^*(scal_g dvol_{g,or}).$$

In particular, the Einstein–Hilbert Lagrangian n-form $scal_g dvol_g$, *where* $dvol_g = dvol_{g,or}$, *is invariant under all orientation preserving diffeomorphisms of* M:

$$scal_{\varphi^*g} dvol_{\varphi^*g} = \varphi^*(scal_g dvol_g).$$

Proof For all diffeomorphisms φ of M we have

$$dvol_{\varphi^*g,\varphi^*or} = \varphi^* dvol_{g,or} \quad \text{and} \quad scal_{\varphi^*g} = \varphi^* scal_g.$$

$\square$

The Einstein–Hilbert Lagrangian can be generalized by considering instead

$$\mathscr{L}(j^2g) = \frac{1}{16\pi\kappa}(scal + 2\Lambda)dvol_g, \tag{5.38}$$

where Λ is a constant known as the *cosmological constant*. The proof of Theorem 5.45 does also show the following.

Theorem 5.47 *The equations of motion of the Einstein–Hilbert Lagrangian with cosmological constant (5.38) are equivalent to*

$$Ric - \frac{1}{2}scal \cdot g = \Lambda g. \tag{5.39}$$

This equation is equivalent to the system

$$\Lambda = \frac{2-n}{2n}scal \quad \text{and} \quad Ric = \frac{scal}{n}g.$$

Notice that if $n \leq 2$, then necessarily $\Lambda = 0$ and every metric is again a solution. If $n \geq 3$, then one can show (exercise) that if g is an *Einstein metric*, that is

$$Ric = fg$$

for some function f, then $f = \frac{scal}{n} = const$. Thus g solves (5.39) for $\Lambda = \frac{2-n}{2n}scal$. In other words, the solutions of (5.39) for $n \geq 3$ are precisely the Einstein metrics such that $scal = \frac{2n}{2-n}\Lambda$. Observe that given an Einstein metric, the equation $scal = \frac{2n}{2-n}\Lambda$ can be always solved by rescaling the Einstein metric by a positive constant provided that $scal$ has the same sign as Λ. So it is sufficient to distinguish only 3 cases: $\Lambda = 0$, $\Lambda > 0$ and $\Lambda < 0$.

To end this section we explain now how to obtain the Einstein metrics as solutions of a variational problem without introducing a cosmological constant. For that we consider the Einstein–Hilbert Lagrangian but restrict to variations g_ε with compact support (that is (5.33) has compact support) that are volume preserving, that is $\frac{\partial}{\partial\varepsilon}\big|_{\varepsilon=0} dvol_{g_\varepsilon} = 0$.

Theorem 5.48 *The equations of motion for the Einstein–Hilbert Lagrangian under volume preserving variations with compact support are equivalent to*

$$Ric^0 = 0, \tag{5.40}$$

where $Ric^0 = Ric - \frac{scal}{n}g$ denotes the trace-free part of Ric. A pseudo-Riemannian metric g is a solution of (5.40) if and only if $Ric = fg$ for some function f.

Proof Owing to (5.34), we obtain (5.40) by projecting (5.39) to its trace-free part.   □

5.5 The Energy-Momentum Tensor

In the previous subsection, we discussed pure gravity, that is, gravity in the absence of matter fields. The key ingredient for coupling matter fields to Einstein gravity is the energy-momentum tensor [13, 14, 19]. It can be introduced by reconsidering two important conservation laws of classical mechanics in the context of classical field theory. Recall that the reason for conservation of energy in classical mechanics is the invariance of the Lagrangian under time-translations. Similarly, invariance under spatial translations implies conservation of momentum.

As a concrete simple example to illustrate at least some of the features of matter-coupled Einstein gravity, let us start by considering a first order field theory for maps $f : \mathfrak{S} \to \mathscr{T}$ from some source manifold $\mathfrak{S}$ to some target manifold $\mathscr{T}$. Since our considerations will be local, we may restrict, by a choice of local coordinates, to $\mathfrak{S} = \mathbb{R}^n$, $\mathscr{T} = \mathbb{R}^m$, and we take the standard volume form on $\mathbb{R}^n$. Then we can consider translations $X = \sum v^\mu \partial_\mu$ in the source manifold $\mathfrak{S} = \mathbb{R}^n$, where v^μ are constants. The vector field X is an infinitesimal automorphism of a Lagrangian $\mathscr{L} \in C^\infty(\mathrm{Jet}^1(\mathbb{R}^n, \mathbb{R}^m))$ if and only if

$$\mathscr{L}(j_x^1 f) = \mathscr{L}(x, f(x), \partial f(x)), \quad \partial f(x) := (\partial_\mu f(x)),$$

does not explicitly depend on $x = (x^\mu) \in \mathbb{R}^n$. This follows from the fact that the prolongation of $X = \sum v^\mu \partial_\mu \in \mathfrak{X}(\mathbb{R}^n)$ is given by $\mathrm{pr}^{(1)}(X) = X \in \mathfrak{X}(\mathrm{Jet}^1(\mathbb{R}^n, \mathbb{R}^m))$. This can be seen either from Lemma 5.13 or by observing that the action of a translation by a vector $v = (v^\mu) \in \mathbb{R}^n$ as a diffeomorphism on $\mathrm{Jet}^1(\mathbb{R}^n, \mathbb{R}^m)$ is simply

$$(x, f(x), \partial f(x)) \mapsto (x + v, f(x), \partial f(x)).$$

The Noether current (5.17) corresponding to the infinitesimal translation $-v$ is given by

$$J^\mu = \sum \partial_\nu f^a v^\nu \frac{\partial \mathscr{L}}{\partial u_\mu^a}(j^1 f) - \mathscr{L}(j^1 f)v^\mu.$$

We can write it as $J^\mu = \sum v^\nu T_\nu^\mu$, where

$$T_\nu^\mu := \partial_\nu f^a \frac{\partial \mathscr{L}}{\partial u_\mu^a}(j^1 f) - \mathscr{L}(j^1 f)\delta_\nu^\mu \tag{5.41}$$

are the components of the *energy-momentum tensor* or *stress-energy tensor*. (Caveat: We will redefine the energy-momentum tensor below.) As a consequence of Noether's theorem (Theorem 5.10) we have the following result.

Proposition 5.49 *If $\mathcal{L} \in C^\infty(\mathrm{Jet}^1(\mathbb{R}^n, \mathbb{R}^m))$ is invariant under translations in $\mathbb{R}^n$, then the energy-momentum tensor is divergence-free (when evaluated on solutions of the equations of motions), that is*

$$\sum \frac{\partial}{\partial x^\mu} T^\mu_\nu = 0. \tag{5.42}$$

Example 5.50 Consider the linear sigma-model

$$\mathcal{L}(j^1 f) = \frac{1}{2} \sum g_{ab}(f) \partial_\mu f^a \partial^\mu f^b - V(f), \quad f : \mathbb{R}^n \to \mathbb{R}^m,$$

with potential $V \in C^\infty(\mathbb{R}^m)$, where greek indices are raised with the inverse of the constant metric $h = \sum h_{\mu\nu} dx^\mu dx^\nu$, for example $\partial^\mu = \sum h^{\mu\nu} \partial_\nu$ and $(h^{\mu\nu})$ being the matrix inverse to $(h_{\mu\nu})$. The corresponding energy-momentum tensor, written as $(0, 2)$-tensor field, is symmetric and given by

$$T_{\mu\nu} = \sum g_{ab}(f) \partial_\mu f^a \partial_\nu f^b - \mathcal{L}(j^1 f) h_{\mu\nu}.$$

In index-free notation the right-hand side reads

$$f^* g - \mathcal{L} h.$$

When $h = dt^2 - \sum_{\alpha=1}^{n-1} (dx^\alpha)^2$ is the Minkowski metric, then the *energy density* (i.e. the charge density associated with the infinitesimal automorphism $-\partial_t$)

$$T^{00} = T^0_0 = \frac{1}{2} g(\partial_t f, \partial_t f) + \underbrace{\frac{1}{2} \sum_{\alpha=1}^{n-1} g(\partial_\alpha f, \partial^\alpha f) + V(f)}_{=: \tilde{V}(f, \partial_1 f, \dots, \partial_{n-1} f)},$$

resembles the energy in mechanics (and coincides with it if $n = 1$), if we consider $\tilde{V}$ as potential energy. The *non-gravitational energy* is in this situation defined as the spatial integral of the energy density:

$$E(t) = \int_{\mathbb{R}^{n-1}} T^{00}(t, x^1, \dots, x^{n-1}) d^{n-1} x. \tag{5.43}$$

It is constant under appropriate boundary conditions at spatial infinity (as in Theorem 5.22). Similarly, one defines the *momentum density* as the time-dependent spatial vector field

$$\sum_{\alpha=1}^{n-1} T^{\alpha 0} \partial_\alpha$$

on $\mathbb{R}^{n-1}$ and the *momentum* as the spatial integral

$$\mathbf{P}(t) = \sum_{\alpha=1}^{n-1} P^\alpha(t)\partial_\alpha, \quad P^\alpha(t) = \int_{\mathbb{R}^{n-1}} T^{\alpha 0}(t, x^1, \ldots, x^{n-1}) d^{n-1}x. \quad (5.44)$$

Notice that the momentum density $(T^{\alpha 0})_{\alpha=1,\ldots,n-1}$ is the flux density associated with the infinitesimal automorphism $-\partial_t$. Its components coincide up to sign with the charge densities $T_\alpha^0 = -T^{\alpha 0}$ associated with the spatial translations $-\partial_\alpha$, $\alpha = 1, \ldots, n-1$.

It is important to remark that the physical notions of energy and momentum as defined above are only valid in the Newtonian limit of Einstein gravity, that is, for quasi-static matter systems in asymptotically flat space-times (see also the discussion in [14, Sect. 19.3]). In general, the physical notions of energy and momentum become rather subtle concepts in Einstein's theory of general relativity. In fact, no general definitions of energy and momentum valid for arbitrary space-time metrics exist. Only for special classes of metrics there exist well-defined expressions such as, for example, ADM energy, Komar energy, Bondi energy and Hawking energy. For further details regarding this subtle aspect of Einstein gravity, the reader is referred to [13, 14, 19].

Example 5.51 Consider the Yang–Mills Lagrangian $\mathscr{L}(j^1 A) = \frac{1}{4}\sum \mathrm{tr}\,(F_{\mu\nu} F^{\mu\nu})$. Differentiating this Lagrangian with respect to $\partial_\mu A_\rho$ (more precisely, with respect to $u^i_{j\rho,\mu}$ in the notation of Sect. 5.4.2) we obtain the (base-point dependent) linear map

$$\mathfrak{g} \to \mathbb{K}, \quad B \mapsto \mathrm{tr}\,(B F^{\mu\rho}).$$

Therefore from (5.41) we obtain that the corresponding energy-momentum tensor is given by

$$T_\nu^\mu(\text{old}) = \sum \mathrm{tr}\,\big((\partial_\nu A_\rho) F^{\mu\rho}\big) - \mathscr{L}(j^1 A)\delta_\nu^\mu.$$

(We are working in a global trivialization of the vector bundle, such that the endomorphisms $F_{\mu\nu}$ can be considered as matrices.) At least in the case of Abelian gauge groups (the reader may try to extend this to the case of non-Abelian gauge groups), this tensor differs from the tensor

$$T_\nu^\mu(\text{new}) = \sum \mathrm{tr}\,(F_{\nu\rho} F^{\mu\rho}) - \mathscr{L}(j^1 A)\delta_\nu^\mu, \quad (5.45)$$

which is symmetric with respect to the metric h, by a tensor with vanishing divergence. In fact, for Abelian gauge groups $F_{\mu\nu} = \partial_\mu A_\nu - \partial_\nu A_\mu$ and it suffices to subtract

$$\sum \text{tr}\left((\partial_\rho A_\nu)F^{\mu\rho}\right) = \sum \partial_\rho \text{tr}\left(A_\nu F^{\mu\rho}\right)$$

from $T_\nu^\mu(\text{old})$. The last equation holds because of the Yang–Mills equation, which for Abelian gauge groups reduces to $\sum \partial_\rho F^{\mu\rho} = 0$. The tensor (5.45), which differs from the original energy-momentum tensor by symmetrization and is still divergence-free, will from now on be called the *energy-momentum tensor* and its components will be denoted by T_ν^μ, rather than $T_\nu^\mu(\text{new})$. Notice that the newly defined energy-momentum tensor is not only coordinate independent (as the previously defined energy-momentum tensor) but, contrary to the previously defined one, is always invariant under gauge transformations.

In the presence of gravity, the space-time metric g (formerly denoted h) is considered as a dynamical field of the theory. The definition of the action functional (5.1) is then modified to read

$$S[f, g] = \int_M L(j^k(f), j^\ell(g))\, dvol_g \,,$$

where $dvol_g = \sqrt{|\det(g)|}\, dx^1 \wedge \cdots \wedge dx^n$ in local coordinates $(x^1, \ldots, x^n)$ on some open set $U \subset M$ (cf. (5.37)). For physical reasons, it is important to distinguish between the gravitational Lagrangian[4] L_{GR} and the matter Lagrangian L_{matter}. The former depends only on g and at least in ordinary Einstein gravity (as opposed to modified gravity theories) it is taken to be the Einstein–Hilbert Lagrangian $L_{EH} = \frac{1}{16\pi\kappa} scal$ (cf. (5.31)). The matter Lagrangian not only depends on g, but also on other fields through their k-th order jets. In total, $L = L_{GR} + L_{matter}$.

One can check that in the above examples the (symmetric) energy-momentum tensor can be obtained from the formula

$$T_{\mu\nu} := \frac{2}{\sqrt{|\det(g)|}} \frac{\delta \mathcal{L}_{matter}}{\delta g^{\mu\nu}} = 2\frac{\delta L_{matter}}{\delta g^{\mu\nu}} + g_{\mu\nu} L_{matter} \,, \tag{5.46}$$

where $\dfrac{\delta \mathcal{L}}{\delta g^{\mu\nu}} := E_{g^{\mu\nu}}(\mathcal{L})$, and

$$E_{g^{\mu\nu}} = \frac{\partial}{\partial g^{\mu\nu}} - \sum \partial_\lambda \frac{\partial}{\partial [\partial_\lambda g^{\mu\nu}]} \pm \cdots \tag{5.47}$$

denotes the Euler–Lagrange operator associated with the field $g^{\mu\nu}$ as defined in (5.3). Here we are denoting the coordinates on the jet space $(u^{\mu\nu}, u_\lambda^{\mu\nu}, \ldots)$ by the same symbols as their evaluation $(g^{\mu\nu}, \partial_\nu g^{\mu\nu}, \ldots)$ on $j^k(g^{-1})$, as customary in the literature. Equation (5.46) is also often used as the definition of the (matter) energy-momentum tensor in the physics literature. With this definition, the energy-momentum tensor is always symmetric.

---

[4]We remark that in some texts the definition of the Lagrangian in gravity theories differs by a factor of $\sqrt{|\det(g)|}$, namely $\mathcal{L} = \sqrt{|\det(g)|}\, L$. This will be used below.

Proposition 5.52 *Let* $\mathcal{L} = \mathcal{L}_{EH} + \mathcal{L}_{matter}$ *be a gravity theory described as the sum of the Einstein–Hilbert Lagrangian for a pseudo-Riemannian metric g and a Lagrangian* $\mathcal{L}_{matter}$ *which depends on g and on other fields through their k-th order jets. Then the equations of motion corresponding to the variation of the metric take the form:*

$$\frac{1}{8\pi\kappa}\left(Ric_g - \frac{1}{2}scal_g\, g\right) = -T, \tag{5.48}$$

where $T = \sum T_{\mu\nu}dx^\mu dx^\nu$ *is the energy-momentum tensor of* $\mathcal{L}_{matter}$ *defined in (5.46).*

Proof The Euler–Lagrange operator associated with $g^{\mu\nu}$ is $\frac{1}{2}T$ for the matter Lagrangian and $\frac{1}{16\pi\kappa}\left(Ric_g - \frac{1}{2}scal_g\, g\right)$ for the Einstein–Hilbert term. For the latter statement we are using the calculation in the proof of Theorem 5.45 and taking into account that the variation of the inverse metric is related to the variation of the metric, $h = \frac{\partial}{\partial\varepsilon}\big|_{\varepsilon=0}g_\varepsilon$, by

$$\frac{\partial}{\partial\varepsilon}\bigg|_{\varepsilon=0} g_\varepsilon^{-1} = -g^{-1}hg^{-1}.$$

□

   Notice that the minus sign on the right-hand side of the general *Einstein equation* (5.48) is usually absorbed by a sign change in the definition of the energy-momentum tensor T.

   The full set of equations of motions corresponding to $\mathcal{L} = \mathcal{L}_{EH} + \mathcal{L}_{matter}$ is the Einstein equation (5.48) together with the matter equations of motions coming from varying $\mathcal{L}$ with respect to the matter fields. Due to the coupling to gravity, the matter equations of motions in general also contain terms involving the metric, thereby turning the full set of equations of motions into a system of coupled equations.

Corollary 5.53 *For every solution of the Einstein equation (5.48), the energy-momentum tensor (5.46) of the matter Lagrangian* $\mathcal{L}_{matter}$ *satisfies the covariant divergence-free condition (cf. Proposition 5.49), namely*

$$\sum_\mu \nabla_\mu T^{\mu\nu} = 0, \tag{5.49}$$

where $T^{\mu\nu} = \sum g^{\mu\rho}g^{\nu\sigma}T_{\rho\sigma}$ *and* ∇_μ *are the components of the Levi-Civita connection of g.*

Proof This follows from the equations of motion provided in Proposition 5.52 and the fact that the *Einstein tensor* $Ric_g - \frac{1}{2}scal_g\, g$ is divergence-free, see [17, Chap. 12, Lemma 2].

□

   Unlike Eq. (5.42), Eq. (5.49) in general does not correspond to a conservation law in the sense of Definition 5.9. The analog of the quantities defined in Eqs. (5.43) and (5.44), that is

$$\int_{\Sigma} \sum \sqrt{|\det(g)|} T^{\mu\nu} d\sigma_{\nu} \,,$$

where the integration is performed over a spacelike hypersurface Σ, is conserved only if $\sum \partial_{\mu}(\sqrt{|\det(g)|} T^{\mu\nu}) = 0$ holds rather than Eq. (5.49) (see, for example, [13, Chap. 11, Sect. 101] for further details). To find a conservation law nevertheless, one needs to also take into account the contribution of the gravitational field (and its derivatives) to the total energy-momentum. This can be achieved by generalizing the definition (5.46) to

$$T^{\text{eff}}_{\mu\nu} := \frac{2}{\sqrt{|\det(g)|}} \frac{\delta \mathscr{L}}{\delta g^{\mu\nu}} = \frac{2}{\sqrt{|\det(g)|}} E_{g^{\mu\nu}}(\mathscr{L}) \,. \tag{5.50}$$

An analysis along the lines of [13, Chap. 11, Sect. 101] shows the following analog of Proposition 5.49 in the presence of gravity:

$$\sum \partial_{\mu} T^{\text{eff}\,\mu\nu} = \sum \partial_{\mu}(T^{\mu\nu} + t^{\mu\nu}) = 0 \,, \tag{5.51}$$

where $T^{\mu\nu} = \sum g^{\mu\rho} g^{\nu\sigma} T_{\rho\sigma}$ is the same as in (5.46) after raising the indices, and $t^{\mu\nu}$ is called the *energy-momentum pseudo-tensor* of the gravitational field. The quantity $t^{\mu\nu}$ is a coordinate-dependent object that cannot be interpreted as the components of a tensor field. Note that Eq. (5.51) is equivalent to Eq. (5.49), with the difference being that the latter is manifestly covariant. From the above, we see that the distinction between energy-momentum carried by the gravitational field versus energy-momentum carried by the matter fields becomes a subtle issue in general relativity. For more details, the reader is referred to [13, 14, 19].

Appendix A
Exercises

A.1 Exercises for Chap. 2

1. Suppose that for every local chart $\varphi : U \to \mathbb{R}^n$ on some smooth manifold M we are given a system of n functions $\alpha_i^\varphi \in C^\infty(U)$, $i = 1, \ldots, n$. Let us denote by $\mathcal{V}_\varphi^i \in C^\infty(U)$ the components of a smooth vector field $\mathcal{V}$ on M with respect to the chart φ. Suppose that for every $\mathcal{V}$ and for every pair of charts $\varphi : U \to \mathbb{R}^n$, $\tilde{\varphi} : \tilde{U} \to \mathbb{R}$ the functions $\sum \alpha_i^\varphi \mathcal{V}_\varphi^i$ and $\sum \alpha_i^{\tilde\varphi} \mathcal{V}_{\tilde\varphi}^i$ coincide on $U \cap \tilde{U}$. Show that there exists a smooth function $f_\mathcal{V}$ on M and a smooth one-form α on M such that $f_\mathcal{V}|_U = \sum \alpha_i^\varphi \mathcal{V}_\varphi^i$ and $\alpha|_U = \sum \alpha_i^\varphi dx^i$ for every chart $\varphi = (x^1, \ldots, x^n) : U \to \mathbb{R}^n$. Check that $\alpha(\mathcal{V}) = f_\mathcal{V}$.
2. Determine the equations of motion of a free particle in Euclidean space (see Example 2.2) and find the general solution.
3. Determine the equations of motion for the harmonic oscillator (see Example 2.2) and find the general solution.
4. Consider the Riemannian manifold $\tilde{M} = (\mathbb{R}^{n+1} \setminus \{0\}, g_{\text{can}})$, where g_{can} is the restriction of the Euclidean metric $\langle \cdot, \cdot \rangle$ on $\mathbb{R}^{n+1}$. Denote by Γ the cyclic group generated by the homothety $x \mapsto 2x$.

   a. Show that the quotient $M = \tilde{M}/\Gamma$ is diffeomorphic to $S^1 \times S^n$ and that the Riemannian metric

   $$\tilde{g} := \frac{1}{r^2} g_{\text{can}}, \quad r(x) := \sqrt{\langle x, x \rangle}, \quad x \in \tilde{M},$$

   induces a Riemannian metric g on M.
   b. Determine all periodic motions for the Lagrangian mechanical system $(M, \mathcal{L})$, where $\mathcal{L}(v) = \frac{1}{2} g(v, v)$, $v \in TM$.
   c. Show that (M, g) is locally conformally flat, that is for every $p \in M$ there exists an open neighborhood $U \subset M$ and a positive function $f \in C^\infty(U)$ such that the Riemannian metric $f \cdot g|_U$ is flat.

© The Author(s) 2017
V. Cortés and A.S. Haupt, *Mathematical Methods of Classical Physics*,
SpringerBriefs in Physics, DOI 10.1007/978-3-319-56463-0

d. Does there exist a global function $f \in C^\infty(M)$ such that fg is of constant (sectional) curvature?

   (*Hint: You may use the classification of complete simply connected Riemannian manifolds of constant sectional curvature. These manifolds are precisely the Euclidean spaces, the spheres and the hyperbolic spaces.*)

5. Let V be a smooth function on a Riemannian manifold (M, g). Consider the Lagrangian

$$L(v) = \frac{1}{2}g(v, v) - V(\pi v), \quad v \in TM, \tag{A.1}$$

where $\pi : TM \to M$ is the canonical projection. Let X be a Killing vector field on (M, g) such that $X(V) = 0$. Check by direct calculation (without using Noether's theorem) that the function $v \mapsto g(v, X(\pi v))$ on TM is an integral of motion of the Lagrangian mechanical system (M, L).

6. Show that the automorphism group of the Lagrangian mechanical system (M, L) of the previous exercise is given by

$$\mathrm{Aut}(M, L) = \{\varphi \in \mathrm{Isom}(M) | V \circ \varphi = V\}.$$

   Deduce that the infinitesimal automorphisms of (M, L) are precisely the Killing vector fields X such that $X(V) = 0$.

7. Let $\gamma : I \to \mathbb{R}^3 \setminus \{0\}$ be a smooth curve. Show that the following conditions are equivalent:

   a. $\gamma \times \gamma' = 0$, where $\times$ denotes the cross product,
   b. γ is a radial curve, that is $\gamma(I)$ is contained in the ray $\mathbb{R}^{>0}v_0$ generated by some constant vector $v_0 \in \mathbb{R}^3 \setminus \{0\}$. Here $\mathbb{R}^{>0}$ denotes the set of positive real numbers.

8. Consider a particle $\gamma : I \to \mathbb{R}^3 \setminus \{0\}$ of mass $m = 1$ moving in Euclidean space under the influence of Newton's gravitational potential $V = -\frac{M}{r}$, where units have been chosen such that the gravitational constant $\kappa = 1$.

   a. Determine the radial motions of the system.
   b. Deduce the total fall time as a function of the initial radius if the initial velocity is zero.

      Hints: Rather than calculating r as a function of time t you will notice that in general it is easier to calculate t as a function of r. The function r of t is then implicitly determined as the inverse function and will not be calculated explicitly. The calculation of $r \mapsto t(r)$ reduces to finding the primitive of a function. For that it might be helpful to calculate the derivative of the function $F(x) = \sqrt{x + x^2} - \mathrm{arsinh}(\sqrt{x})$ for $x > 0$.

9. (Conservation of angular momentum). Consider a particle $t \mapsto \gamma(t) \in \mathbb{R}^3$ moving in Euclidean space according to Newton's law

$$\frac{d}{dt}\mathbf{p} = F.$$

Its *angular momentum* at time t is the vector

$$\mathbf{L}(t) := \gamma(t) \times \mathbf{p}(t).$$

Show that $\mathbf{L}$ is subject to the equation

$$\frac{d}{dt}\mathbf{L} = \mathbf{M},$$

where $\mathbf{M} := \gamma \times F$ is the *moment of force*. Deduce that the angular momentum is constant if the moment of force is zero.

10. Let V be a smooth function on a pseudo-Riemannian manifold (M, g) and consider the Lagrangian $\mathcal{L}(v) = \frac{1}{2}g(v, v) - V(\pi v)$, $v \in TM$. Assume that with respect to some coordinate system on M the metric g and the potential V are both invariant under rotations in one of the coordinate planes, say in the (x^1, x^2)-plane. Show that there exists a corresponding integral of motion defined on the coordinate domain. How is it related with the notion of angular momentum?

11. Show that a function on $\mathbb{R}^3 \setminus \{0\}$ is radial if and only if it is *spherically symmetric*, that is invariant under $SO(3)$. Deduce that invariance under $SO(3)$ is equivalent to invariance under $O(3)$.

12. Show that a vector field on $M = \mathbb{R}^3 \setminus \{0\}$ is radial if and only if it is *spherically symmetric*, that is invariant under the natural action of $SO(3) \subset \mathrm{Diff}(M)$. Deduce that invariance under $SO(3)$ is equivalent to invariance under $O(3)$.
    Hint: Recall that the natural action of the diffeomorphism group $\mathrm{Diff}(M)$ on the vector space of vector fields $F : M \to TM$ on a smooth manifold M is given by

$$F \mapsto F^{\varphi} := d\varphi \circ F \circ \varphi^{-1}, \quad \varphi \in \mathrm{Diff}(M).$$

13. Let $\gamma : I \to \mathbb{R}^3 \setminus \{0\}$ be the motion of a particle in a radial force field F according to Newton's law $m\gamma'' = F(\gamma)$. Recall that the motion is planar due to the conservation of the angular momentum vector. Show that the area $A(t_0, t_1)$ swept out by the vector γ during a time interval $[t_0, t_1]$ is given by $A(t_0, t_1) = \frac{L}{2m}(t_1 - t_0)$.

14. Let (M, g) be a pseudo-Riemannian manifold and f a nowhere vanishing smooth function on M. Consider the pseudo-Riemannian metric

$$g_N = g + f\,du^2$$

on $N := M \times \mathbb{R}$, where u is the coordinate on $\mathbb{R}$. Show that geodesic equations for a curve $t \mapsto \gamma_N(t) = (\gamma(t), u(t)) \in N = M \times \mathbb{R}$ can be separated into

a. the equations of motion for γ with respect to a Lagrangian of the form $\mathcal{L}(v) = \frac{1}{2}g(v, v) - V(\pi v)$, $v \in TM$, where V is a certain smooth function on M related to f and $\pi : TM \to M$ is the canonical projection and

    b. an ordinary differential equation for the function $t \to u(t)$, which can be solved by integration once we know $t \mapsto \gamma(t)$.

15. Determine the radius r as a function of the angle φ for a motion of a particle of unit mass with non-zero angular momentum in Coulomb's electrostatic potential. Can you use the results about the motion in Newton's gravitational potential? What is the main difference?

16. Let V be a smooth function on a pseudo-Riemannian manifold, which is either bounded from above or from below. Show[1] that there exists a function f related to V and a pseudo-Riemannian metric $g_N = g + f du^2$ on $N = M \times \mathbb{R}$, where u is the coordinate on the $\mathbb{R}$-factor, such that the solutions of the Lagrangian mechanical system defined by $\mathscr{L}(v) = \frac{1}{2}g(v, v) - V(\pi v), v \in TM$, correspond precisely to geodesics $t \mapsto \gamma_N(t) = (\gamma(t), u(t))$ in (N, g_N) with a particular choice of the affine parameter t and which satisfy $u' \neq 0$.

17. Let V be a smooth function, which is bounded from below, on a complete Riemannian manifold (M, g). Show that the solutions of the Lagrangian mechanical system defined by $\mathscr{L}(v) = \frac{1}{2}g(v, v) - V(\pi v), v \in TM$, exist for all times. *Hint: You may use the previous exercise to relate the problem to the completeness of a Riemannian manifold (N, g_N) of the type $N = M \times \mathbb{R}$, $g_N = g + f du^2$. Recall that a Riemannian metric on a smooth manifold M is called complete if every Cauchy sequence in (M, g) converges and that this notion is equivalent to geodesic completeness by the Hopf–Rinow theorem.*

18. Deduce from the previous exercise that for every smooth function V on a compact Riemannian manifold the solutions of the Lagrangian mechanical system defined by $\mathscr{L}(v) = \frac{1}{2}g(v, v) - V(\pi v), v \in TM$, exist for all times.

A.2 Exercises for Chap. 3

19. Let f be a smooth function on a symplectic manifold M. Compute the Hamiltonian vector field X_f in a coordinate system $(q^1, \ldots, q^n, p_1, \ldots, p_n)$ defined on some open subset $U \subset M$ such that $\omega|_U = \sum dp_i \wedge dq^i$.

20. Let $(M, \mathscr{L})$ be a Lagrangian mechanical system and denote by $\pi : TM \to M$ the canonical projection. Show that the following conditions are equivalent:

    a. $\mathscr{L}$ is non-degenerate.
    b. $\phi_{\mathscr{L}} : TM \to T^*M$ is a local diffeomorphism.
    c. For all $x \in M$, $\phi_{\mathscr{L}}|_{T_xM} : T_xM \to T_x^*M$ is of maximal rank.
    d. For all $v \in TM$, there exists a coordinate system (x^i) defined on an open neighborhood U of πv such that the matrix $\left(\frac{\partial^2 \mathscr{L}(v)}{\partial \hat{q}^i \partial \hat{q}^j}\right)$ is invertible, where $(q^1, \ldots, q^n, \hat{q}^1, \ldots, \hat{q}^n)$ are the corresponding coordinates on TU.

---

[1] See [6] for results generalizing this exercise and also Exercise 14.

e. For all $v \in TM$, and every coordinate system (x^i) defined on an open neighborhood U of πv the matrix $\left(\frac{\partial^2 \mathscr{L}(v)}{\partial \dot{q}^i \partial \dot{q}^j}\right)$ is invertible.

21. Let $(M, \mathscr{L})$ be a Lagrangian mechanical system. Show that the following conditions are equivalent:

   a. $\mathscr{L}$ is nice.
   b. $\phi_\mathscr{L} : TM \to T^*M$ is a diffeomorphism.
   c. $\mathscr{L}$ is non-degenerate and for all $x \in M$, $\phi_\mathscr{L}|_{T_xM} : T_xM \to T_x^*M$ is a bijection.
   d. For all $x \in M$, $\phi_\mathscr{L}|_{T_xM} : T_xM \to T_x^*M$ is a diffeomorphism.

22. Let (M, g) be a pseudo-Riemannian manifold and denote by $\phi : TM \to T^*M$ the isomorphism of vector bundles induced by g. Let V be a smooth function on M and consider the Lagrangian $\mathscr{L}(v) = \frac{1}{2}g(v, v) - V(\pi v)$, $v \in TM$. Here $\pi : TM \to M$ denotes the projection. Denote by $E \in C^\infty(TM)$ the energy and by $H = E \circ \phi^{-1} \in C^\infty(T^*M)$ the Hamiltonian.

   a. Show that if a curve $\tilde{\gamma} : I \to T^*M$ is a motion of the Hamiltonian system (T^*M, ω, H), then the curve $\pi \circ \tilde{\gamma} : I \to M$ is a motion of the Lagrangian system $(M, \mathscr{L})$. Here ω denotes the canonical symplectic form.
   b. Show that the map $\gamma \mapsto \phi \circ \gamma'$ from curves in M to curves in T^*M is inverse to the map $\tilde{\gamma} \mapsto \pi \circ \tilde{\gamma}$ from curves in T^*M to curves in M when restricted to solutions of the Euler–Lagrange equations and Hamilton's equations, respectively. Here $\pi : T^*M \to M$ denotes the projection.

23. State and prove a local result relating Lagrangian mechanical systems with non-degenerate Lagrangian to Hamiltonian systems, similar to the global result proven in Sect. 3.2.2 for nice Lagrangians.

24. Let f be a smooth function on a finite-dimensional real vector space V such that $\phi_f : V \to V^*$ is a diffeomorphism and consider its Legendre transform $\tilde{f} \in C^\infty(V^*)$. Show that $\phi_{\tilde{f}} : V^* \to V$ is also a diffeomorphism and that the Legendre transform of $\tilde{f}$ is f.

25. A subspace $U \subset V$ of a symplectic vector space (V, ω) is called *isotropic* if $U \subset U^\perp$. Show that the maximal dimension of an isotropic subspace is $n = \frac{1}{2} \dim V$ and that the isotropic subspaces U of dimension n are Lagrangian, that is $U = U^\perp$. Deduce that, with the definitions given in Chap. 4, an immersed submanifold of a symplectic manifold is Lagrangian if and only if its tangent spaces are Lagrangian.

26. Let (T^*M, H) be a Hamiltonian system of cotangent type, $n = \dim M$, and let $S : M \times U \to \mathbb{R}$ be a smooth n-parameter family of solutions of the Hamilton–Jacobi equation. Show that the following conditions are equivalent.

   (i) The family is non-degenerate.
   (ii) The map $\Phi_S : M \times U \to T^*M$, defined in Definition 4.7, is a local diffeomorphism.
   (iii) For all $x \in M$,

$$\Phi_S|_{\{x\}\times U} : \{x\} \times U \cong U \to T_x^*M, \quad u \mapsto dS^u|_x,$$

is a local diffeomorphism.

(iv) For all $x \in M$,

$$\Phi_S|_{\{x\}\times U} : \{x\} \times U \cong U \to T_x^*M,$$

is of maximal rank.

(v) For all $(x, u) \in M \times U$ the $n \times n$-matrix

$$\left(\frac{\partial^2 S(x, u)}{\partial x^i \partial u^j} \right)$$

is invertible, where (x^i) are local coordinates in a neighborhood of $x \in M$ and (u^i) are (for instance) standard coordinates in $U \subset \mathbb{R}^n$.

27. Let (M, g) be a Riemannian manifold and let $x_0 \in M$ be an equilibrium point of a given Lagrangian mechanical system $\mathcal{L}(v) = \frac{1}{2}g(v, v) - V(\pi v)$. Further assume that x_0 is a local minimum and a non-degenerate critical point of V. The latter means that the Hessian of V with respect to the Levi-Civita connection at x_0 is positive definite. Show that x_0 is a stable solution in the sense of Lyapunov stability, i.e. show that $\forall \varepsilon > 0 \ \exists \delta > 0$, such that for all solutions $\gamma : I \to M$, $0 \in I$, of the corresponding Euler–Lagrange equations the following holds true:

$$\sqrt{d_M(x_0, \gamma(0))^2 + g_{\gamma(0)}(\dot{\gamma}(0), \dot{\gamma}(0))} < \delta$$

$$\Rightarrow \sqrt{d_M(x_0, \gamma(t))^2 + g_{\gamma(t)}(\dot{\gamma}(t), \dot{\gamma}(t))} < \varepsilon \qquad \forall t \in I,$$

where $d_M(\cdot, \cdot)$ denotes the metric on M induced by g.

Hints: Use the Morse lemma to bring V near x_0 to a certain form. Verify that for $K \subset M$ compact and contained in a chart domain, one can find $c > 0$ and $C > 0$, such that $c\langle \cdot, \cdot \rangle \leq g|_K(\cdot, \cdot) \leq C\langle \cdot, \cdot \rangle$, where $\langle \hat{q}, \hat{q} \rangle = \sum \hat{q}_i^2$. The topology of M coincides with the topology induced by d_M, see [11, p. 166], in particular V is continuous with respect to the metric d_M. Also, recall that the energy is an integral of motion, hence estimating $E(\dot{\gamma}(0))$ yields a global result for γ.

28. a. Consider a Lagrangian mechanical system in $\mathbb{R}^3$ with radial potential $\mathcal{L}(v) = \frac{1}{2}\langle v, v \rangle - V(r)$ and assume that $r(t) = r_0$ is the constant radial component of a motion with $|\mathbf{L}| > 0$. Recall the corresponding expression for the energy

$$E = \frac{1}{2}\dot{r}^2 + V_{\text{eff}}(r)$$

and show that such a solution is stable in the sense of the previous exercise if

$$\frac{d^2V}{dr^2}(r_0) + \frac{3}{r_0}\frac{dV}{dr}(r_0) > 0.$$

b. Consider $V(r) = -\frac{\alpha}{r^n}$, $\alpha > 0$. Find all $n \in \mathbb{N}$ such that a stable motion $r(t) = r_0$ exists. Does the answer depend on α? (Note that this question is related to Newtonian gravitation in $d = n + 2$ dimensions. To show this, one needs to use a conservation law similar to the angular momentum which was used in dimension three.)

c. For which $b > 0$ does the Lagrangian system corresponding to the potential $V(r) = -\frac{\alpha}{r}\exp\left(-\frac{r}{b}\right)$, $\alpha > 0$, have stable motions with $r(t) = r_0$? Does the answer depend on α?

d. For which α does the Lagrangian system corresponding to the potential $V(r) = \alpha \ln(r)$ have stable motions with $r(t) = r_0$?

29. Let $q^1, \ldots, q^n, \hat{q}^1, \ldots, \hat{q}^n$ denote the standard coordinates on $T\mathbb{R}^n$. Consider the function $V : \mathbb{R} \setminus \{0\} \to \mathbb{R}$ defined by

$$V(x) = V_0\left(\left(\frac{x_0}{x}\right)^{12} - \left(\frac{x_0}{x}\right)^{6}\right),$$

where $V_0, x_0 \in \mathbb{R} \setminus \{0\}$ are constants. The Lagrangian system

$$\mathcal{L} = \frac{1}{2}\langle\hat{q}, \hat{q}\rangle - \sum_{i=1}^{n-1} V(q^{i+1} - q^i)$$

describes small oscillations in a linear chain of n atoms of unit mass. Note that the oscillations are assumed to be in direction of the chain.

a. Determine the distance a of two neighboring atoms in the equilibrium position.

b. Show that the Taylor expansion up to second order of V at a is given by

$$\tilde{V}(x) = -\frac{V_0}{4} + \frac{1}{2}k(x - a)^2$$

with $k = \frac{18}{\sqrt[3]{2}}\frac{V_0}{x_0^2}$.

c. Determine the equations of motion corresponding to the linearized problem

$$\mathcal{L}_2 = \frac{1}{2}\langle\hat{q}, \hat{q}\rangle - \sum_{i=1}^{n-1} \tilde{V}(q^{i+1} - q^i).$$

A.3   Exercises for Chap. 4

30. Consider a point q on a conic section $C \subset \mathbb{R}^2$ with focal points $f_1 \neq f_2$. Show that the tangent line at q bisects the angle formed by the two lines connecting q with the focal points.

31. Consider a hyperbola $H \subset \mathbb{R}^2$ with focal points f_1, f_2 at distance $2c > 0$. Recall that H is defined by the equation $|r_1 - r_2| = 2a$, where r_1 and r_2 are the distances to f_1 and f_2, and $a \in (0, c)$ is a constant. Show that the restriction of the Euclidean metric to H is given by

$$\frac{\xi^2 - 4a^2}{4(\xi^2 - 4c^2)} d\xi^2,$$

where $\xi = r_1 + r_2$. (This result is used in the proof of Proposition 4.11.)

A.4   Exercises for Chap. 5

32. Consider the Lagrangian $\mathscr{L}(q, \hat{q}) = \prod_{i=1}^n \hat{q}^i$ on $\mathbb{R}^n$ and let $n \geq 2$. Show that the restriction of $\phi_{\mathscr{L}} : T\mathbb{R}^n \to T^*\mathbb{R}^n$ to the open subset

$$U_\sigma = \{(q, \hat{q}) \in T\mathbb{R}^n \mid \sigma\mathscr{L} > 0\} \subset T\mathbb{R}^n,$$

where $\sigma \in \{+, -\}$, is a diffeomorphism onto its image. Check that $\phi_{\mathscr{L}}(U_+) = \phi_{\mathscr{L}}(U_-)$ if n is odd and that $\phi_{\mathscr{L}}(U_+)$ and $\phi_{\mathscr{L}}(U_-)$ are invariant under the antipodal map in the fibers of $T^*\mathbb{R}^n$ if n is even. Determine the corresponding Hamiltonians on $\phi_{\mathscr{L}}(U_\sigma)$ and find the general solution of the corresponding Hamilton's equations.

33. Show that
$$\dim \mathrm{Jet}_0^k(\mathbb{R}^n, \mathbb{R}) = \binom{n+k}{k}.$$

34. Show that for every smooth manifold there are canonical identifications

   a.
   $$\mathrm{Jet}^1(M, \mathbb{R}) = \mathbb{R} \times T^*M,$$

   b.
   $$\mathrm{Jet}^1(\mathbb{R}, M) = \mathbb{R} \times TM.$$

35. Let $P = (P^1, \ldots, P^n)$ be a smooth vector-valued function on $\mathrm{Jet}^k(\mathbb{R}^n, \mathbb{R}^m)$. Its total divergence $\mathrm{Div}\, P$ is a smooth function on $\mathrm{Jet}^{k+1}(\mathbb{R}^n, \mathbb{R}^m)$. We know from Proposition 5.5 that the Euler operators E_a, acting on functions on

$\text{Jet}^{k+1}(\mathbb{R}^n, \mathbb{R}^m)$, vanish on Div P. Check this in the following cases by directly computing $E_a(\text{Div } P)$:

a. $k = 0$,
b. $k = m = n = 1$.

36. a. Show that the Euler–Lagrange equations of the free scalar field Lagrangian on the n-dimensional Minkowski space $(M, g) = (\mathbb{R}^n, (dx^0)^2 - \sum_{i=1}^{n-1}(dx^i)^2)$, given in index notation by

$$\mathscr{L} = \frac{1}{2}\partial_\mu \varphi \partial^\mu \varphi - \frac{1}{2}m^2\varphi^2,$$

yield the Klein–Gordon equation $(\Box + m^2)\varphi = 0$.

b. Verify that in the above case $\Box = (-1)^{n-1} \star d \star d$, where $\star$ denotes the Hodge star-operator, defined by

$$\alpha \wedge \star\beta = \langle \alpha, \beta \rangle dvol,$$

where α, β are differential forms on n-dimensional Minkowski space and $dvol$ is the metric volume form of the Minkowski metric $(dx^0)^2 - \sum_{i=1}^{n-1}(dx^i)^2$ compatible with the standard orientation.

37. Consider the Lagrangian of the Kepler problem with Newton's potential in $\mathbb{R}^3$,

$$\mathscr{L}(v) = \frac{1}{2}\langle v, v \rangle + \frac{M}{r}.$$

Show that each entry of the so-called Runge-Lenz vector (or Laplace-Runge-Lenz vector)

$$R := \begin{pmatrix} x_t \\ y_t \\ z_t \end{pmatrix} \times \left(\begin{pmatrix} x \\ y \\ z \end{pmatrix} \times \begin{pmatrix} x_t \\ y_t \\ z_t \end{pmatrix} \right) - \frac{M}{r}\begin{pmatrix} x \\ y \\ z \end{pmatrix}$$

is an integral of motion.

38. For a given $k \in \mathbb{N}$, determine the Euler–Lagrange equations for

$$\mathscr{L} = \frac{1}{2}\sum_{|J| \leq k} u_J^2, \quad \mathscr{L} \in C^\infty(\text{Jet}^k(\mathbb{R}^n, \mathbb{R})).$$

39. Let (M, g) and (N, h) be pseudo-Riemannian manifolds of dimension m and n, respectively. We assume that (N, h) is oriented and denote its volume form by $dvol_h$. Determine the Euler–Lagrange equations for the Lagrangian n-form $\mathscr{L} dvol_h$ defined by

$$\mathscr{L}(j^1 f) = \frac{1}{2}\langle df, df \rangle, \quad f \in C^\infty(N, M),$$

where $\langle \cdot, \cdot \rangle_x$ is the scalar product on $T_x^* N \otimes T_{f(x)} M$, $x \in N$, induced by h and g. Show that the resulting equations are equivalent to

$$\mathrm{tr}_h \nabla df = 0,$$

where ∇ is the connection on the vector bundle $T^* N \otimes f^* TM \to N$ induced by the Levi-Civita connections on M and N.

40. Let X and Y be smooth vector fields on $\mathbb{R}^n$ and $\mathbb{R}^m$, respectively, $Z = X + Y$, $dvol$ the standard volume form of $\mathbb{R}^n$, and $\mathscr{L} \in C^\infty(\mathrm{Jet}^k(\mathbb{R}^n, \mathbb{R}^m))$ a Lagrangian. Assume that

$$(\mathrm{pr}^{(k)} Z)(\mathscr{L}) + \mathscr{L} \mathrm{div} X$$

is a total divergence. Show that the same calculation as in the proof of Noether's theorem can be used to prove that

$$\sum Q^a E_a(\mathscr{L}), \quad \text{defined by} \quad Q^a = Y^a - \sum u_i^a X^i,$$

is a total divergence, where X^i and Y^a are the components of $X \in \mathfrak{X}(\mathbb{R}^n)$ and $Y \in \mathfrak{X}(\mathbb{R}^m)$, respectively.

41. Let (M, g) and (N, h) be pseudo-Riemannian manifolds and $V \in C^\infty(N \times M)$. Consider the Lagrangian

$$\mathscr{L}(j^1 f) = \frac{1}{2} \langle df, df \rangle - V(j^0 f), \quad f \in C^\infty(N, M).$$

   a. In the special case when (N, h) is a pseudo-Euclidean vector space, show that $\mathscr{L} dvol_h$ is invariant under the group of translations of the source manifold N if $V \in C^\infty(M) \subset C^\infty(N \times M)$ and compute the corresponding Noether currents.
   b. In the special case when (M, g) is a pseudo-Euclidean vector space, show that $\mathscr{L} dvol_h$ is invariant under the group of translations of the target manifold M if $V \in C^\infty(N) \subset C^\infty(N \times M)$ and compute the corresponding Noether currents.

42. Let $\Omega \subset \mathbb{R}^{n-1}$ be a bounded domain with smooth boundary. We consider a Lagrangian $\mathscr{L} \in C^\infty(\mathrm{Jet}^k(U, \mathbb{R}^m))$ on the cylinder $U = \mathbb{R} \times \Omega \subset \mathbb{R}^n$ with standard coordinates $(x^0, x^1, \ldots, x^{n-1}) =: (t, \mathbf{x})$. Let $P : \mathrm{Jet}^\ell(\mathbb{R}^n, \mathbb{R}^m) \to \mathbb{R}^n$ be a smooth vector-valued function such that

$$\mathrm{Div}\, P|_{\mathrm{Jet}^{\ell+1}(U, \mathbb{R}^m)}$$

is a conservation law of $\mathscr{L} dvol$, where $dvol$ is the standard volume form of U. Let $f \in C^\infty(U, \mathbb{R}^m)$ be a solution of the Euler–Lagrange equations which extends smoothly to a neighborhood of the closure of U. We decompose the current $J = P(j^\ell f) = (J^0, \mathbf{J}) : U \to \mathbb{R}^n$ into the charge density J^0 and the flux density **J**. Show that the time evolution of the charge $Q(t) = \int_\Omega J^0(t, \mathbf{x}) d^{n-1}x$ is given by

$$Q'(t) = -\int_{\partial\Omega} \langle \mathbf{J}, \nu \rangle dvol_{\partial\Omega},$$

where ν stands for the outer normal of $\partial\Omega$ and $dvol_{\partial\Omega}$ for the induced volume form on $\partial\Omega \subset \mathbb{R}^{n-1}$. Conclude that the charge is constant if the flux density is tangent along the boundary of Ω.

43. Consider Yang–Mills theory on a Hermitian line bundle over Minkowski space $(M, g) = (\mathbb{R}^4, dt^2 - \sum_{\alpha=1}^3 (dx^\alpha)^2)$. Since the Lie algebra of U(1) is one-dimensional, we can identify the curvature $F = F^\nabla$ with an ordinary real-valued 2-form. Show that the Yang–Mills equation $d * F = 0$ reduces to half of Maxwell's equations in the vacuum, that is

$$\operatorname{div} \mathbf{E} = 0, \quad \operatorname{rot} \mathbf{B} = \frac{\partial}{\partial t}\mathbf{E},$$

whereas the Bianchi identity $dF = 0$ reduces to the other half of Maxwell's vacuum equations, that is

$$\operatorname{div} \mathbf{B} = 0, \quad \operatorname{rot} \mathbf{E} = -\frac{\partial}{\partial t}\mathbf{B},$$

where $\mathbf{E} = \sum E^\alpha \partial_\alpha$ and $\mathbf{B} = \sum B^\alpha \partial_\alpha$ are the time-dependent vector fields on $\mathbb{R}^3$ obtained by decomposing F into

$$E^\alpha = F(\partial_t, \partial_\alpha) \quad \text{and} \quad B^\alpha = -F(\partial_\beta, \partial_\gamma),$$

where (α, β, γ) runs trough cyclic permutations of $\{1, 2, 3\}$.

44. Let ∇ be a G-connection in a vector bundle E and α a one-form with values in $\mathfrak{g}(E)$. Show that the curvature of the G-connection $\nabla + \alpha$ is given by

$$F^{\nabla+\alpha} = F^\nabla + d^\nabla\alpha + \alpha \wedge \alpha.$$

45. Express the Yang–Mills equations in local coordinates.

46. Check by direct calculation that the Noether current associated with an infinitesimal gauge transformation as stated in Proposition 5.44 is conserved, that is divergence-free.

47. Prove that

$$\frac{\partial}{\partial\varepsilon}\bigg|_{\varepsilon=0} dvol_{g_\varepsilon} = \operatorname{tr}(g^{-1}h)dvol_g,$$

for every smooth family of pseudo-Riemannian metrics g_ε, where $g = g_0$ and $h = \frac{\partial}{\partial\varepsilon}\big|_{\varepsilon=0} g_\varepsilon$.

References

1. R. Abraham, J.E. Marsden, *Foundations of Mechanics* (Benjamin Cummings Publishing Company, Reading, 1978)
2. V.I. Arnold, *Mathematical Methods of Classical Mechanics* (Springer, Berlin, 1984)
3. S. Bates, A. Weinstein, Lectures on the geometry of quantization. Berkeley Math. Lect. Notes **8**, 1–137 (1997)
4. S.K. Donaldson, P.B. Kronheimer, *The Geometry of Four-Manifolds* (Oxford University Press, New York, 1990)
5. A. Einstein, *The Meaning of Relativity* (Princeton University Press, Princeton, 2005)
6. L.P. Eisenhart, Dynamical trajectories and geodesics. Ann. Math. **30**, 591–606 (1929)
7. H. Goldstein, *Classical Mechanics*, 2nd edn. (Addison-Wesley, Menlo Park, 1980)
8. H. Goldstein, C. Poole, J. Safko, *Classical Mechanics*, 3rd edn. (Addison-Wesley, Menlo Park, 2002)
9. J.D. Jackson, *Classical Electrodynamics*, 2nd edn. (Wiley, New York, 1975)
10. W. Klingenberg, *Lineare Algebra und Geometrie* (Springer, Berlin, 1984)
11. S. Kobayashi, K. Nomizu, *Foundations of Differential Geometry*, vol. 1 (Wiley, New York, 1996)
12. B. Kosyakov, *Introduction to the Classical Theory of Particles and Fields* (Springer, Berlin, 2007)
13. L.D. Landau, E.F. Lifshitz, *The Classical Theory of Fields* (Butterworth-Heinemann, London, 1980)
14. C.W. Misner, K.S. Thorne, J.A. Wheeler, *Gravitation* (W. H. Freeman, San Francisco, 1973)
15. K. Moriyasu, *An Elementary Primer for Gauge Theory* (World Scientific Publishing Company, Times Mirror, 1983)
16. P.J. Olver, *Applications of Lie Groups to Differential Equations* (Springer, Berlin, 1993)
17. B. O'Neill, *Semi-Riemannian Geometry* (Academic Press, Dublin, 1993)
18. J.J. Sakurai, *Modern Quantum Mechanics* (Benjamin Cummings Publishing Company, Reading, 1985)
19. R.M. Wald, *General Relativity* (University of Chicago Press, Chicago, 1984)

© The Author(s) 2017
V. Cortés and A.S. Haupt, *Mathematical Methods of Classical Physics*,
SpringerBriefs in Physics, DOI 10.1007/978-3-319-56463-0

Index

© The Author(s) 2017
V. Cortés and A.S. Haupt, *Mathematical Methods of Classical Physics*,
SpringerBriefs in Physics, DOI 10.1007/978-3-319-56463-0

Printed in the United States
By Bookmasters